THE IMPORTANCE OF TIME

PHILOSOPHICAL STUDIES SERIES

VOLUME 87

THE IMPORTANCE OF TIME

Proceedings of the Philosophy of Time Society, 1995–2000

Edited by

L. NATHAN OAKLANDER
The University of Michigan-Flint,
Flint, Michigan, U.S.A.

KLUWER ACADEMIC PUBLISHERS
DORDRECHT / BOSTON / LONDON

A C.I.P. Catalogue record for this book is available from the Library of Congress.

ISBN 978-90-481-5841-6

Published by Kluwer Academic Publishers,
P.O. Box 17, 3300 AA Dordrecht, The Netherlands.

Sold and distributed in North, Central and South America
by Kluwer Academic Publishers,
101 Philip Drive, Norwell, MA 02061, U.S.A.

In all other countries, sold and distributed
by Kluwer Academic Publishers,
P.O. Box 322, 3300 AH Dordrecht, The Netherlands.

Printed on acid-free paper

Printed in the Netherlands

To my sister, Dianne Hope Oaklander, my aunts,
Helen Rosenberg and Rhoda Winters, and the memory
of my uncles Jack Rosenberg and Leonard Winters

TABLE OF CONTENTS

CONTRIBUTORS

WILLIAM LANE CRAIG , Research Professor of Philosophy, Talbot School of Theology
MARK HINCHLIFF, Associate Professor of Philosophy, Reed College.
RONALD C. HOY, Professor and Chair of Philosophy, California University of Pennsylvania
DAVID P. HUNT, Professor of Philosophy, Whittier College
BRIAN LEFTOW, Professor of Philosophy, Fordham University
ROBIN LE POIDEVIN, Professor of Metaphysics, University of Leeds.
NED MARKOSIAN, Associate Professor of Philosophy, Western Washington University
IAIN MARTEL, Instructor of Philosophy, University of Colorado at Boulder
STORRS MCCALL, Professor of Philosophy, McGill University
D. H. MELLOR, Emeritus Professor of Philosophy, University of Cambridge
L. NATHAN OAKLANDER, Professor of Philosophy, University of Michigan-Flint
L. A. PAUL, Assistant Professor of Philosophy, University of Arizona; Research Fellow, Research School of Social Sciences, Australian National University
JOHN PERRY, Henry Waldgrave Stuart Professor of Philosophy, Stanford University
ERIC M. RUBENSTEIN, Assistant Professor of Philosophy, Indiana University of Pennsylvania
QUENTIN SMITH, Professor of Philosophy, Western Michigan University
MICHAEL TOOLEY, Professor of Philosophy, University of Colorado at Boulder

PREFACE

The Philosophy of Time Society grew out of a National Endowment for the Humanities Summer Seminar on the Philosophy of Time offered by George Schlesinger in 1991. The members of that seminar wanted to promote interest in the philosophy of time and Jon N. Turgerson offered to become the first Director of the society with the initial costs underwritten by the Drake University Center for the Humanities. Thus, the Philosophy of Time Society (PTS) was formed in 1993. Its goal is to promote the study of the philosophy of time from a broad analytic perspective, and to provide a forum as an affiliated group with the American Philosophical Association, to discuss the issues in and related to the philosophy of time.

The society held its first meeting during the Eastern Division of the APA in Atlanta, George, in December 1993. In 1997 I began my tenure as Executive Director of PTS and with my term ending in 2000, I decided to put together a volume of selected papers read at PTS meetings over the years. The result is the present volume. It contains some of the latest developments in the field, including discussions of recent books by Michael Tooley, *Time, Tense, and Causation*, and D. H. Mellor, *Real Time II*, and much more. The main issue in the philosophy of time is and remains the status of temporal becoming and the passage of time. Is the "movement" of time from the future to the present and into the past to be understood in terms of some fundamental and irreducible change that a moment of time or an event in time undergoes when it "changes" from being future to being present to being past? Or are temporal becoming and the passage of time devoid of any real ontological significance reflecting only our different temporal perspective from which events or moments of time are viewed? This issue, however fundamental, is not the only issue that the papers in this volume deal with. In fact, one of the reasons for the resurgence of interest in the question of temporal becoming is the close connection between the main issue or temporal becoming and so many other issues that philosophers have come to see are intimately connected with it. The papers in this volume reflect and contribute to understanding the importance of the philosophy of time to other fundamental issues of perennial concern. For example, getting clear on the different aspects of the A- and B-theory debate can shed light on other debates in areas such as the philosophy of mind, the philosophy of science, the philosophy of language, the philosophy of religion, and metaphysics. Individuals who are interested in causation, the problem of change, Frank Jackson's knowledge argument, divine foreknowledge, the phenomenology of time, time and physics, time and physical objects, God and time, Parmenides and time, as well as arguments for and against A and B-theories of time, will find much that is of interest in this book.

Most of the papers in this volume follow the format in which they were presented: lead symposia and colloquia papers are followed by a commentator's reply. Thus, two or more sides of an issue are presented and defended, and it is left to the reader to pursue the truth.

L. Nathan Oaklander

Flint,
Michigan

ACKNOWLEDGEMENTS

Some of the papers first presented at PTS meetings where subsequently published and I thank the editors for permission to reprint the following material: Ronald C. Hoy, "Parmenides Complete Rejection of Time," *Journal of Philosophy* XCI, 11 (November 1994): 573-598. David Hunt, "Two Problems with Knowing the Future," *American Philosophical Quarterly* 34 (April 1997): 273-285. Ned Markosian, "What are Physical Objects?" *Philosophy and Phenomenological Research*, 61 (2000): 375-395, and L. Nathan Oaklander, "Review of *Time, Tense, and Causation* by Michael Tooley," *Mind*, 108, 430 (April 1999): 407-413. I also wish to thank my secretary, Sue Fabbro, for her excellent work in preparing the camera-ready copy of this manuscript, and my wife Linda for her continued encouragement, patience and support. Funding for this book was partially supported by the University of Michigan-Flint.

Part One

BOOK SYMPOSIA

1
TOOLEY ON TIME AND TENSE

L. Nathan Oaklander
University of Michigan-Flint

Since the publication of Richard Gale's anthology on *The Philosophy of Time* (1967), and his original work, *The Language of Time* (1968), and especially since D. H. Mellor's *Real Time* (1981), the literature on the philosophy of time has flourished.[1] Part of the reason for this renewed interest in the philosophy of time concerns its connection with many other issues in the philosophy of religion, the philosophy of language, the philosophy of science and metaphysics. In Michael Tooley's case, it was his research on the nature of causation that led him to a consideration of the nature of time and the debate between the dynamic (or tensed) and the static (or tenseless) views of time. In his book, *Causation: A Realist Approach* (1987) Tooley argued for a causal analysis of the direction of time and *in Time, Tense, and Causation* (1997) he argues that a full account of the relation between causation and time is even deeper since events can be causally related only in a dynamic world; a world in which the past and present are real, but the future is not. The version of the "non-existent future" theory of time that Tooley endorses is original and in various ways quite unlike more traditional tensed theories. He espouses some theses that have hitherto been rejected by tensers, for example, "that tenseless concepts and facts are more basic than tensed ones" (1997, p. 29), but he argues for them very carefully.

This is an important book that makes a valiant effort to synthesize the insights of both tensers and detensers. In it Tooley has many interesting things to say on a wide variety of topics such as backward causation, indeterminism, counterfactuals, truth-functionality and the logical connectives, the analysis of tensed statements, McTaggart's paradox, absolute and relational theories of space and time (or space-time), the Special Theory of Relativity, and Storrs McCall's model of a dynamic world, to name just a few. In the confines of a limited review I cannot possibly do justice to the richness and sophistication of the numerous arguments in this book. Rather, I shall limit myself to summarizing the main argument for his version of the dynamic view as it unfolds throughout the book and indicating what I believe are the most serious difficulties facing it.

L. N. Oaklander (ed.), The Importance of Time, 3–12.

In Part I, "Causation, Time and Ontology," Tooley states his basic ontological position and gives his central argument – the argument from causation – in support of it. Tooley's basic position may be summarized as follows. Traditionally the dispute between tensers and detensers has centered around two issues. First, is the world static or dynamic? That is, does change involve an object (or the world as a whole) having different properties at different times, as the tenseless theory contends, or is there change in an object (or the world) if and only if "the totality of temporal facts, or states of affairs, [constituting an object or the world as a whole] is different at different times" (1997, p. 14), as the tensed theory maintains. On the static conception, what states of affairs are actual does not depend upon what time it is, whereas on the dynamic view it does.[2] The second issue is whether tensed or tenseless facts are basic? As Tooley puts it, are tensed facts logically supervenient on tenseless facts, as the B-theorist believes, or are tenseless facts logically supervenient on tensed facts, as traditional A-theorists assert? Tooley's novel thesis is that an adequate theory of time requires a dynamic world in which the *basic* facts are *tenseless*. His theory, therefore, combines what he takes to be the essential core of both A and B-theories; the flow of time implies the coming into existence at a time of basic tenseless facts. Time passes as new tenseless states of affairs are added to the sum total of what exists (tenselessly) as of a specified time.

In order to support this dynamic theory Tooley accepts as primitive the notions of actual at a time (and the corresponding notion of truth at a time), and actual *simpliciter* (and the corresponding notion of truth *simpliciter*). He needs the former notion to account for the change over time in what states of affairs exist, and he needs the latter to account for (*inter alia*) the meaning of statements that are about or involve the future (1997, pp. 40-41 and 152-158).

Given this machinery, Tooley's argument for the dynamic view has the following structure. Tooley's conclusion is that the proper analysis of the nature of causation implies that the world is a dynamic one in which the past and present are real, but the future is not. He supports that conclusion by first arguing against reductionist approaches to causation. Next he claims that the concept of causation cannot be analytically basic since causation is not given in immediate experience. As reductionist accounts have been ruled out "the only possibility, then, is to treat causation as a theoretical relation" (1997, p. 99). Furthermore, since Tooley argues for a singularist conception of causation (1997, pp. 95-98) he cannot characterize the relation of causation by reference to causal laws. Thus, he formulates a theory of causation according to which "[C]ausation is that unique relation between events or states of affairs such that, if there are laws involving that relation, those laws must satisfy postulates (C_1) to (C_4)" (1997, p. 103). These postulates concern the relation between posterior probabilities and prior probabilities and they assert that the relation is different for causes than for effects. The idea is that if a law satisfies those postulates then it is a causal law, and if it is causal law then it involves a causal relation.

Tooley then argues that causality implies the dynamic view: The existence of causal laws must satisfy certain postulates concerning the transmission of possibilities. In particular they must satisfy the postulates (C_4) and (C_1) that imply "the posterior probability of an effect depends upon the prior probability of the cause...whereas...the posterior probability of a cause does not depend upon the prior probability of its effect" (1997, pp. 107-08). On the static view, or on a dynamic view according to which past, present and future are all equally real or in which only the present exists, there will be no way to capture the asymmetry involved in causation. On the other hand, if there is a deep ontological asymmetry in the world, where the past and present do exist and the future does not, then the causal asymmetry expressed by the key postulates can be explained. As Tooley puts it, "the asymmetry that is expressed, in particular, by postulates (C_1) and (C_4) will be captured and explained" only if "causal laws, in conjunction with what is actual as of a given time, determine what states of affairs will be added to what is already actual, and, thereby, what *will ultimately exist*, in a *tenseless sense*" (1997, p. 111; my emphasis).

Before we can accept this argument for Tooley's version of the dynamic view we need to understand what the view actually holds. In particular, what does it mean to say that a state of affairs "will exist (tenselessly)." For something to exist tenselessly does not mean that it did, does, or will exist, since Tooley takes the existential quantifier to be tenseless (1997, p. 175). Thus, if an event E or state of affairs exists *tenselessly*, then it exists *simpliciter*, without regard to what time it is, or the concepts of past, present or future. What analysis, then, can be given of the statement "Event E will exist (tenselessly)?" If E exists tenselessly, then it exists *simpliciter*, but if it is future, then on Tooley's view, it does not exist as of the present moment. It seems to me that either this view reduces to the tenseless theory, or it involves a contradiction. Although Tooley does offer an analysis of ordinary tensed sentences, he never really avoids this basic dilemma, as I shall demonstrate more fully below.

In Part II, "Semantical Issues," Tooley argues that his theory requires two concepts of truth: Truth at a time and truth *simpliciter*. Since what tenseless facts there are changes with the passage of time, there must be the idea of truth at a time, and since no tensed account can be satisfactory unless it incorporates the idea of the totality of existence, there must be the concept of truth *simpliciter* which in the case of tenseless propositions can be "recursively defined [as] correspondence between propositions, and the sum total of what exists" (1997, p.153). In recognizing two kinds of truth, truth at a time and truth *simpliciter*, and in recognizing two kinds of existence, existence at a time and existence *simpliciter*, Tooley wants to have his cake and eat it too. On the one hand, he wants reality to consist of "the mereological union of *all* the states of affairs that are actual as of one time or another" (1997, p. 153; my emphasis). That is, he wants reality to be the totality of what (tenselessly) exists *simpliciter*. On the other hand, to allow for the accretion of facts, his view presupposes that the sum total of what exists (tenselessly) is different at different times. Unfortunately, Tooley never clearly explains how the sum total of existence can remain the same and have temporal parts that change their existential status. Tooley asserts that the problem of change motivates the debate between tensed and tenseless approaches to time (1997, p. 13), but he does not realize that his own solution simply reraises it. He maintains that the totality exists *simpliciter* and yet new tenseless facts come into existence and are added on to the totality as time passes. But how is this possible? How can the totality exist *simpliciter* and remain the same through a change (as of different times) in the existential status of its temporal parts?

In Part III, "Tensed Facts," Tooley is concerned with the meaning of tensed language. First, he argues that although the concept of temporal relations is not analytically basic, the various accounts by tensers to analyze tenseless temporal concepts in terms of tensed ones are all unsuccessful. Then he argues that while ordinary tensed statements, for example, "It will rain," do involve indexicals and thus express different propositions at different times, there are more basic tensed statements, for example, "Event E lies in the future at time t," that do not contain indexicals. Thus, for example, "Event E lies in the future at time t" is more basic than an ordinary tensed statement because the meaning, in the sense of truth conditions, of a non-indexical tensed statement can be used to give the meaning of an ordinary tensed sentence. On his analysis, "Event E is future at time t" means

> Event E is later than time t, and t is an instantaneous state of affairs, t is actual as of time t, and no state of affairs that is later than t is actual as of time t. (1997, p. 196)

The problem with this analysis is this: If Event E is later than time t, and temporal priority is a tenseless concept, then E is later than time t is a state of affairs that (tenselessly) exists *simpliciter*. But then how can it be the case that "no state of affairs that is later than t is actual as of time t," unless Tooley simply means that no state of affairs later than t is located at t? And if that is what he means, then his theory is compatible with the *tenseless* conception of time.

Chapter 8 contains criticisms of alternative tensed accounts of time and a defense of the most common objections to the tenseless theory. (He takes the argument from causation to be a decisive refutation of tenseless time.) Tooley considers those views that hold that events have one or more intrinsic non-relational tensed properties, and argues that they fail since, among other things, they cannot explain how an *instantaneous* event can have an A-property at one time and then lose it at *another* time. On the other hand, if the future is not real then, according to Tooley, tensed properties can be viewed as relational properties of events and the difficulty is avoided since, "[A]n instantaneous event may, therefore, possess the relational property of pastness at one time, and the relational property of presentness at another time" (1997, p. 229). Tooley also offers some penetrating criticisms of (two versions of) presentism and of Storrs McCall's version the dynamic view.

In Part IV, "Temporal Relations," Tooley puts forth and defends a novel version of the causal theory of time. According to Tooley, qualitative tenseless temporal relations and the direction of time are necessarily identical with causal relations, whereas a modification has to be made for quantitative temporal relations. Before turning to those accounts Tooley attempts to disarm a potential difficulty with the causal theory. He notes that if one is concerned with making sense of our ordinary notion of time, which is absolute, then it would seem that one couldn't hold a causal theory of time. For if temporal relations were analyzable in terms of causal relations between states of affairs then it would seem that in a world devoid of events and objects causal relations cannot exist. This, Tooley argues, is not the case since there is no *a priori* reason for denying that spatiotemporal regions can enter into causal relations, and there appear to be good reasons for accepting a substantival view of space-time (1997, p. 261). He then proceeds with his analysis of the qualitative temporal relations of simultaneity and priority, which he gives in terms of actual causal relations plus spatial relations. He then analyses quantitative temporal relations, i.e., relations that are a matter of the temporal distance between events, and that relation can be defined as that unique relation among events and/or space-time points "whose involvement in causal laws is required if the laws are to entail the existence of potential correlations between relevant causal processes" (1997, p. 279). Tooley concludes this chapter with a consideration of objections to causal theories. Since the objections are typically based on a relational theory of time his version of the causal theory is immune to them. The second sort of objections is to absolute space and time. Tooley argues that these arguments are either based on a discredited verificationism or the assumption of a Humean supervenience thesis that he rejects.

In Part V, "Objections," Tooley considers both philosophical and scientific objections to his version of the dynamic view. One serious philosophical argument is that put forth by McTaggart. Tooley thinks that it is a non-starter since he holds that (a) temporal relations are not analyzable in terms of tensed properties, and that (b) the alleged incompatible properties problem is avoided by maintaining that events are past, present, and future, neither simultaneously nor timelessly, but at different tenseless times. Thus, Event E is future at time t and that is compatible with Event E is present at time t_1 and Event E is past at t_2. However, I don't think that McTaggart's paradox can be dealt with so easily. Tooley maintains that statements about the future, though not true at the present moment, may be true *simpliciter* (1997, p. 303). However, if for example, "Clinton will leave office in the year 2000" is true *simpliciter* then, given his commitment to the correspondence theory of truth, there must be some corresponding state of affairs that is actual *simpliciter*. If there is such a state of affairs, then it would appear to be none other than the state of affairs that comes into existence as of the year 2000. However, if that state of affairs exists *simpliciter* and also comes into existence at a certain time, then one and the same state of affairs both exists and does not exist. To avoid the contradiction by asserting that these are two different primitive notions of existence strikes me as obfuscating.

Tooley devotes Chapter eleven to the scientific objection to the dynamic view based upon the Special Theory of Relativity. Put simply the objection is that on the Special Theory all simultaneous events are such relative to a frame of reference. On the dynamic theory, where the (non-existent) future is later than the present, there must be some absolute, frame independent, relation of simultaneity such that no event which is future relative to one frame is present relative to another. Tooley does not follow Stein's (1968) gambit or any of the other recent responses to this difficulty. Rather, he carefully defends a new strategy which is to modify the Special Theory of Relativity so that it does not entail that the one way speed of light is a constant in all directions in all inertial frames, but does allow for a definition of absolute simultaneity. He then offers three reasons for believing that the modified theory is preferable to the Special Theory. First, its predictive and explanatory power is greater. Second, there is no empirical evidence in support of the one way principle. And finally, results in quantum mechanics, specifically the collapse of the wave pocket, support the view that simultaneity is absolute.

Part VI, "Summing Up," contains a very useful overview of the central features of Tooley's approach to time. What is clear, in this chapter and throughout the book, is that Tooley wants to follow an intermediate path between tenseless approaches to the nature of time and traditional tensed accounts. I do not, however, think that he succeeds, because his view collapses into one or the other of the theories he seeks to avoid. I shall conclude this review by explaining the reasons for my misgivings concerning Tooley's dynamic view of time.

Tooley claims to be defending a tensed theory, but he rejects several theses typically associated with the A-theory and accepts several theses typically associated with the B-theory. For example, Tooley rejects the presentist view that only the present exists – that the past and future are species of unreality – he rejects the view that temporal relations are analyzable in terms of temporal properties, and he rejects primitive and irreducible tensed properties and tensed facts. On the other hand, he maintains "that all of the [ontologically] basic facts are *tenseless* ones" (1997, p. 186) and "that there can be a complete and consistent description of the world that is not a description from any particular temporal perspective" (1997, p. 16). The existence of a complete and consistent description of the world is a consequence of his thesis that there must be the notion of the totality of reality that exists *simpliciter*. Since these are all theses upheld by proponents of the *tenseless* theory, we are left to wonder what then is it that makes Tooley's theory a *tensed* theory of time?

Tooley claims that "[T]he fundamental thing that separates tensed and tenseless accounts of the nature of time is, ... the acceptance or rejection, of a dynamic conception of the world" (1997, p. 16). However, his interpretation of the dynamic aspect of time makes it difficult to understand how his view differs from the B-theory or, if it does, how it avoids the alleged pitfalls of the traditional tensed theory. Tooley says that the key feature of the dynamic view is that "while the past and present are real, the future is not." As of the present moment the future is nothing, a non-entity, but as time passes "the world changes [or grows] through the addition of new *tenseless* facts" (1997, p. 153; my emphasis). Thus, "[I]n order for the world to be a dynamic one all that is required is that the facts that are actual as of one time differ from the facts that are actual as of some other time" (1997, p. 19). However, this last statement is ambiguous. It may mean that the tenseless states of affairs are located (tenselessly or *simpliciter*) at one time are different from the tenseless states of affairs that are located at some earlier or later time. A thesis that is compatible with the tenseless theory of time. Or, it could mean, what I believe Tooley intends, namely, that the sum total of existence at one time differs from the sum total of existence at another time. That is, the totality of reality contains more at a later time t_2 than it does at an earlier time t_1; a thesis compatible with the tensed theory. In either case, the crucial question is, How can Tooley believe *both* that there is a totality of tenseless states of affairs, that neither did, do or will exist, but exist *simpliciter*, and that tenseless facts that are not actual as of one time can become actual as of a later time? To get at the difficulty in reconciling these two theses we need to revisit his notion of the totality of existence.

Tooley's rationale for the concept of the totality of existence is as follows:

> ...besides being able to speak of what is actual as of the present moment, one needs also to be able to make sense of propositions about what was actual, and about what will be actual, and more generally about the totality of what is actual at some time or another. In short, one needs to be able to offer an account of the concept of a total, dynamic world, as contrasted with the history of a dynamic world up to some point in time. (1997, p. 40)

But is not the very concept of a "total, dynamic world" a contradiction in terms? How can the world be a totality of existence and be dynamic? Is it not absurd to suppose that in addition to tenselessly existing at the time they do, there is some further change that tenseless states of affairs undergo "when" they becoming actual or come into existence? Indeed, what sense can be made of the idea that new *tenseless* facts get added to what exists *simpliciter*? If the totality exists *simpliciter* then there is (tenselessly) nothing that can be added to them. Growth could only be something that happens to the series that tenselessly exists, but Tooley gives no explanation of what such a happening could be. It cannot be that the terms of the series successively acquire and shed the property of *presentness* since there is no such property in Tooley's ontology. Nor could the terms of the totality acquire new relations to states of affairs as they come into existence since all the terms of the totality exist *simpliciter* with the properties and relations they have. Thus, the coming into existence of tenseless states of affairs must be something extra, something added onto temporal reality, something that is the myth of passage.

The problem with Tooley's theory can be stated still differently. Tooley claims that we must

> distinguish between being actual as of a given time, and being actual *simpliciter*. *The future is part of the totality of reality, and so is actual-simpliciter*. But the future is not actual as of the present moment, and it is the latter proposition that one is expressing -- or should be expressing -- when it is said that the future is not real. (1997, p. 303; emphasis added)

On the one hand, the claim that "the future is part of the totality of reality, and so is actual *simpliciter*" is perfectly compatible with the tenseless view since it implies that all states of affairs, including those which are "future" exist tenselessly. Moreover, if to say that "the future is not actual as of the present moment" means that events that are not located (tenselessly) at one time are located (tenselessly) at a later time, then that too is compatible with the B-theory. On the other hand, if he means that as of the present moment the future is excluded from the sum total of reality, then this contradicts his thesis that the future exists *simpliciter*.

The same difficulty reemerges when we consider tensed sentences plus the notion of truth. He asserts that we want to make claims such as

(1) It will be true that there are (tenselessly) unicorns, which entails that
(2) There will be unicorns, and from (2) it follows that
(3a) It is (tenselessly) true that there will be unicorns (1997, p. 304).

One question that arises is: What is the truth condition of the tenseless truth that there will be unicorns? What is the structure of the tenseless fact to which it corresponds? Whatever it is, when viewed from the point of view of the totality of existence, the future exists, and statements about the future are either true or false *simpliciter*. When viewed from the perspective of the dynamic theory, future events do not exist, and statements about the future have an indeterminate truth value. But how can, say, "Gore will be President" be true or false *simpliciter*, and also have an indeterminate truth value as of October 7, 1997? If it is true or false *simpliciter*, then it must be so in virtue of some tenseless fact to which it corresponds, but if such a fact exists *simpliciter*, then how can it fail to make true the judgment made on October 7, 1997 that "Gore will be president?" I don't think that there is a consistent set of answers to these vexing questions.

I conclude, therefore, that Tooley's view does not succeed in combining the tenseless and tensed approach, but collapses into either a purely tenseless theory, or a tensed theory that is, in spite of his assertions to the contrary, susceptible to the dialectical difficulties of the traditional version. Nevertheless, my criticisms should not belie the fact that Tooley's provocative book makes an extremely valuable contribution to the literature on the philosophy of time and is mandatory reading for anyone working in the area.[3]

NOTES

[1] Particularly important is Craig (2000a), (2000b), and (2000c), Le Poidevin (1992) and (1998), Lucas (1989), Ludlow (1999), McCall (1994), Mellor (1998), Oaklander (1984), Oaklander and Smith (1994), Schlesinger (1980), Smart (1981), and Smith (1993).

[2] Characterizing B-time as "static," as Tooley and many other A-theorists do, is prejudicial since it suggests that on the tenseless theory nothing really changes, which is, of course, at the center of the debate between tensed and tenseless theories.
[3]I wish to thank Hugh Mellor and Quentin Smith for their comments on an earlier draft of this essay.

REFERENCES

Craig, W. L. (2000a), *The Tenseless Theory of Time: A Critical Examination* (Dordrecht: Kluwer).

Craig, W. L. (2000b), *The Tensed Theory of Time: A Critical Examination* (Dordrecht: Kluwer).

Craig, W. L. (2000c), *Time and the Metaphysics of Relativity* (Dordrecht: Kluwer).

Gale, R. (1968), *The Language of Time* (New York: Humanities Press).

Hinchliff, M. (1996), "The Puzzle of Change," in J. E. Tomberline (ed.), *Philosophical Perspectives* 10: *Metaphysics* (Oxford: Blackwell): 119-136.

Le Poidevin, R. (1998), *Questions of Time and Tense* (Oxford: Clarendon Press).

Le Poidevin, R. (1992), *Change, Cause and Contradiction* (New York: St. Martin's Press).

Lucas, J. R. (1989), *The Future* (Oxford: Blackwell).

Ludlow, P. (1999), *Semantics, Tense, and Time: An Essay in the Metaphysics of Natural Language* (Cambridge, MA: MIT Press).

McCall, S. (1994), *A Model of the Universe* (Oxford: Clarendon Press).

Mellor, D. H. (1998), *Real Time II* (London: Routledge).

Mellor, D. H. (1981), *Real Time* (New York: Cambridge University Press).

Oaklander, L. N. (1984), *Temporal Relations and Temporal Becoming*: A Defense of a Russellian Theory of Time (Lanham, MD: University Press of America).

Oaklander, L. N. and Smith, Q. (eds.), (1994), *The New Theory of Time* (New Haven: Yale University Press).

Prior, A. N. (1968), *Time and Tense* (Oxford: Oxford University Press).

Schlesinger, G. (1980), *Aspects of Time* (Indianapolis: Hackett Publishing Company).

Smart, J. J. C. (1981), "Time and Becoming," in Peter van Inwagen (ed.), *Time and Cause*, (Boston: Reidel): 1-15.

Smith, Q. (1993), *Language and Time* (New York: Oxford University Press).

Stein, H. (1968), "On Einstein-Minkowski Space-Time," *The Journal of Philosophy* 65: 5-23.

Tooley, M. (1986), *Causation: A Realist Approach* (Oxford: Oxford University Press).

Tooley, M. (1997), *Time, Tense, and Causation* (Oxford: Oxford University Press).

2
TOOLEY ON TIME

Storrs McCall
McGill University

Michael Tooley has written a most interesting book on time *(Time, Tense, and Causation*, 1997). In it he argues (i) that to understand time requires that one adopt the idea of a dynamic world, and (ii) that the choice between a tensed and a tenseless theory of facts and of truth should not be an either/or, but a both/and. That is, we must be prepared to accept both the notions of "truth at a time" and "truth *simpliciter,*" of "existence or actuality at a time" and "existence or actuality *simpliciter*" (1997, pp. 40, 154, 303, 375, 381). Can Tooley have his cake and eat it too? Can he consistently combine the idea that we live in a world that is dynamic – that as he puts it what tenseless facts there are depends upon what time it is – with the notion of truth *simpliciter*?

I believe that he can, and that the criticisms of his position made by Nathan Oaklander can be answered. I also believe that the particular conception of a dynamic world that Tooley adopts is not the most satisfactory, and that a better one can be devised. But this is a detail. In general Tooley is right in holding that a dynamic conception of time can and must be combined with a timeless conception of truth, i.e. truth *simpliciter*. This shall be the theme of my talk.

To clarify what is meant by a "dynamic" view of time I shall make a terminological suggestion. Tooley, and other authors, frequently refer to a "tensed" as opposed to a "tenseless" conception of time, to "tensed" vs. "tenseless" facts, to "tenseless truth," etc. Strictly speaking it is sentences and propositions, not time or truth or events, that are either tensed or tenseless. In what follows I shall restrict the distinction tensed/tenseless to sentences, utterances and propositions. Time itself, and truth, and the world of events are neither tensed nor untensed. For time and the world, the contrasts are between "dynamic" and "static," not "tensed" and "tenseless." In the case of truth, the contrast is between "truth at a time," and "timeless" or "atemporal" or "omnitemporal" truth, i.e. "truth *simpliciter.*"

In using the words "dynamic" and "static," care is called for. Something that is "dynamic" is something that is different at different times: contrast a dynamic with a static exchange rate. In the case of the dynamic view of

L. N. Oaklander (ed.), The Importance of Time, 13–19.

time the thing that would have to be different at different times could only be the world as a whole, all of it including past, present and future. But how could this be? The world as a whole is constituted by all the objects and events that are, were or will be, each located in a specific region of space-time. How could this totality be different at different times?

The question being asked is not the question of how the totality of objects and events located at time *t* could differ from the totality of objects and located at time *t'*. In the 4-dimensional Minkowski world, which is the prototypal embodiment of a "static" conception of time, there will in general be enormous differences between a 3-dimensional cross-section of the 4-dimensional world at *t* and a similar cross-section of the world at *t'*. No, the question being asked is how the totality of *all* objects and events at *all* spatio-temporal locations can differ at two different times *t* and *t'*.

Phrased in this way, the short answer might seem to be, it can't. To philosophers accustomed to think of the world 4-dimensionally the very asking of the question might seem to betray a deep misunderstanding of time and of space-time. One of the virtues of Tooley's *Time, Tense and Causation* is that it contains a clear and consistent description of a 4-dimensional world which does differ, as a whole, from one time to another. This is C.D. Broad's "no future" theory, according to which the world consists entirely of past and present things, but not of future things. The latter have no existence. The difference between the world at an earlier and at a later date is that at the later time a quantity of objects and events have been *added* to the world, objects and events that had literally no existence at the earlier time. In Broad's words "A moment *t* is later than a moment *t'* if the sum total of existence at *t* includes the sum total of existence at *t'* together with something more" (1923, p. 67, quoted by Tooley at p. 171). We may call this the "accretional" view of the world; it is a dynamic view which holds that the future is non-existent, and that the world grows by the continuous addition of new objects and events. When they are added these objects and events are present, but thereafter they form part of the past. Broad later, in 1959, writes of "the continual supersession of what was the latest phase by a new phase, which will in turn be superseded by a new phase," and refers to this quality as "the rock-bottom peculiarity of time" (1959, p. 766).

So far so good. Tooley gives a clear account of the accretional theory, and the world to which the accretional theory applies is clearly a dynamic world, which in turn serves as the basis for a dynamic conception of time. The "flow" of time (a word which Tooley never uses, but which I shall dare to use at this point) is constituted by the progressive accretion of objects and events at the "top" end of the 4-dimensional world. Dedicated four-dimensionalists in the tradition of Reichenbach, Carnap, Grünbaum and

Smart may roll their eyes at this point, but the accretional theory provides a perfectly consistent model of a dynamic world.

But wait. The accretional theory may appear to yield a consistent model of a dynamic world, but how does it sit with the concept of truth? Does truth *simpliciter* conflict with the dynamism of the accretional model? It is at this point that Oaklander's criticisms of Tooley's book become relevant. Tooley distinguishes between "true at a time" and "true *simpliciter*" (1997, pp. 148-154), as well as between "actual as of a time" and "actual *simpliciter*" (1997, p. 303). Ignoring the question of logical truth, on which Tooley's views are somewhat obscure, a factual tenseless proposition is *true at a time* if it corresponds to the world as it is as of a given time (1997, p. 137). (In personal communication, Tooley explains that what exists or is actual *at* a given time includes only those things or events with time coordinate *t*, whereas what exists or is actual *as of t* includes everything with time coordinate less than or equal to *t*.) In contrast, a factual tenseless proposition is *true simpliciter* if it corresponds to "the sum total of what exists" (1997, p. 153), or "existence or actuality *simpliciter*" (1997, p. 40), or "the total, dynamic world, as contrasted with the history of a dynamic world up to some point in time" (1997, p. 40), or "the totality of all states of affairs" (1997, p. 377).

Consider for example the proposition

(1) The earth collides (tenselessly) with a 50-foot meteorite in the year 2050.

As of 1998 (1) is not true, and is not false, since the world up to 1998 contains neither a truth-maker nor a falsehood-maker for it. Instead, Tooley says, (1) takes Lukasiewicz's third truth-value (1997, pp. 129-130). This being said, however, Tooley also holds that although (1) may lack both truth and falsehood as of 1998, nevertheless it is either true *simpliciter* or false *simpliciter*. The reason is that when "the totality of all states of affairs" is considered, not just the totality as of 1998, then either a meteorite's hitting the earth in 2050 is included in that totality or it is not. If it is, then (1) is true *simpliciter*; if it is not then (1) is false *simpliciter*.

For Oaklander, Tooley's position on this matter is inconsistent. If (1) is true *simpliciter* there must be a corresponding state of affairs, i.e. an event of the year 2050, that is actual *simpliciter*. But if (1) is not true as of 1998, but becomes true as of 2050, then it does so because of some state of affairs that is *not* actual before 2050. How, Oaklander asks, can there be a state of affairs that makes (1) true *simpliciter*, or true at all times including 1998, but which does not become actual until 2050? It would seem that one and the same state of affairs must both exist and not exist (Oaklander, Essay 1, p. 6).

Tooley appears to have painted himself into a corner, and must answer Oaklander. He can do so, I believe, in the following way, while still retaining the essence of what he wants to say about a dynamic world, and about truth.

Consider the following mundane example:

(2) Oaklander *ties* (tenselessly) his shoelace at noon tomorrow, where the reader can replace "tomorrow" by the appropriate date.

Lacking a truth- or falsehood-maker for (2) today, Tooley would no doubt consider the truth-at-a-time-value of (2) to be neither true nor false.

Tooley could, though he did not, have taken Ducasse's advice (1997, p. 124) and maintained that any qualification of the adjectives "true" and "false" by a date, or for that matter by a place, is absurd. According to Ducasse, "*p* is true on Friday March 27" makes no more sense than "*p* is true in Winnipeg." However, Tooley wishes to sustain both the notion of truth at a time and the notion of truth *simpliciter*. The truth-at-a-time-value of (2), today, is neither true nor false. Even though he holds that tomorrow's events, being in the future, literally do not exist, and will not be added to the growing body of events located in space-time until tomorrow, can Tooley still maintain that (2) has a truth-value *simpliciter*? He can do so without inconsistency, I believe, provided he takes the relationship that holds between truth and events to be an asymmetric relationship of dependence. Let me elaborate.

Truth depends on events, not events on truth. This seems almost a truism, but it is the key to understanding truth *simpliciter*. Truth and falsehood supervene upon events: they and the propositions which are their bearers "float above" events in space-time, being affected without affecting them. Furthermore truth and falsehood *simpliciter* are atemporal: a true tenseless proposition containing no indexicals has the property that if anyone utters a token of it at any time, that person makes a true statement.

Applying all this to (2), when noon tomorrow comes there will be an event of one of the following two kinds: (i) Oaklander's stooping and tying his shoelace, or (ii) Oaklander's not doing so. Neither of those events yet belongs (in Tooley's dynamic ontology) to the space-time manifold of events, but at noon tomorrow one of them will. If it's (i), then (2) is true *simpliciter*; if it's (ii) then (2) is false. Since (2) is true or false atemporally, it's true or false whether uttered in 1000 BC or 3000 AD, or for that matter today. We don't *know* whether it's true or false, but that's irrelevant. In this way Tooley can combine the accretional ontology, which serves as the basis for his dynamic conception of time, with the notion of truth *simpliciter* for tenseless propositions. No contradiction results.

I wish to conclude by focussing more narrowly on Tooley's dynamic ontology, the "no future" ontology of C. D. Broad. In chapter 11 Tooley discusses an obvious weakness, namely that it does not conform to the principles of special relativity. The moment at which events get added to space-time to form a continuously enlarging manifold is the world-wide moment simultaneous with "now," and Tooley maintains that the points belonging to this moment are *absolutely* simultaneous, not merely simultaneous with "now" in some coordinate frame or other. This contradicts the relativity of simultaneity and the equivalence of inertial frames, which lies at the core of Einstein's theory.

Tooley argues valiantly that the idea of a *preferred* frame, a frame that is ontologically distinguished from all the rest, and that determines unequivocally which events get added to space-time at each instant, may represent an improvement upon the STR. But I am afraid that few will be convinced. Relativity theory is confirmed in so many experimental situations, and 4-dimensional Minkowski geometry in itself is such an elegant and beautiful theory, that no one would willingly give it up unless an extremely attractive alternative were there to take its place. What for example would Tooley do about the twins paradox? Would they still be the same age when they met again? On pp. 356-360 Tooley argues that giving up special relativity and adopting absolute simultaneity plays a role in understanding the EPR experiment. Let's consider this idea.

Suppose that a pair of electrons is created in an entangled spin state. If, using a Stern-Gerlach measuring instrument, the spin of one of the electrons is measured along any axis and found to be spin-up (+), the spin of the other electron measured along the same axis will always be spin-down (-). It does not matter which component of spin of the electrons is measured. As long as the axes of the two measuring instruments are parallel, the two electrons will always be found to have opposite spins. Call this the "principle of zero total spin."

One of the many interesting questions that can be asked about the EPR experiment is whether the two instruments are measuring spin values (+ or -) that the electrons *already* have before they are measured or whether the electrons *acquire* determinate spin values along the measured axis at some point during the experiment. Bell showed that if the spin values were determined by local hidden variables or instruction kits that accompanied each electron on its way to the Stern-Gerlach magnets, then the observed outcome statistics over many trials, using different angles of alignment and misalignment of the axes of the two magnets, would satisfy the so-called Bell inequalities. But in experimental situations the Bell inequalities are violated. Hence the EPR measuring instruments are measuring spin values that the electrons somehow acquire during the experiment.

This raises the next interesting question, namely when? Does the left electron, for example, acquire its spin value at the moment it is measured, or at the moment the right electron is measured, or at some other time?

Suppose we denote the left measurement event by "A," and the right measurement event by "B." Suppose that the left instrument measures x-spin, i.e. spin along the x-axis. Let the time of A be *t*, and let B be later than A. Then, plausibly, the left electron acquires determinate x-spin at *t*, when it is measured, and (Tooley argues) the right electron must also acquire determinate x-spin at *t*, before it is measured. If there were a temporal gap between the moment the left electron acquired determinate spin and the moment the right electron did, then according to Tooley an x-spin measurement applied to the right electron after A and before the right electron acquired determinate x-spin could yield results which violated the principle of zero total spin.

To avoid any possibility of violating this principle, Tooley maintains that there can never be any temporal gap between the instant at which one electron acquires determinate spin and the instant at which the other does. But for this condition to be met, Tooley maintains, absolute simultaneity is required. Since the two electrons are moving in opposite directions, there will always be a frame of reference in which the times at which the electrons acquire determinate spin are simultaneous. But equally, there will also be frames in which one electron acquires determinate spin before the other. In that case a temporal gap opens up in which, Tooley says, observed violations of the principle of zero total spin might be found. He concludes that the principles of special relativity cannot apply to the EPR experiment, and that we must adopt a privileged frame and re-install absolute simultaneity.

This is not so, however. One of the principles of quantum mechanics, at least in its traditional Copenhagen form, is that definite values of physical quantities such as spin cannot be assigned to quantum systems unless a "measurement" of that quantity is made. If, to go back to Tooley's argument, there were a temporal gap between the moment at which the left electron acquired determinate spin and the moment at which the right electron did, there could still be no violation of the principle of zero total spin unless a measurement of the right electron's spin were made. And of course, if the left electron's x-spin were "spin-up," then the right electron's x-spin would always be measured to be "spin-down." This holds no matter what temporal gap there may be between the measurements.

The situation described in the last paragraph would be the case no matter which frame of reference we adopted: a frame (i) in which A occurs before B, or a frame (ii) in which B occurs before A, or a frame (iii) in which A and B are simultaneous. (Since A and B will in general be space-like separated events, lying outside each other's light cones, there will always be frames of these three types.) But no matter which of (i)-(iii) we choose, the outcomes of the x-spin measurements will always be opposite, and no violation of the principle of zero total spin will be observed. Absolute simultaneity is not required, and the relativistic principle of the equivalence of inertial frames can be maintained.

In conclusion, one shortcoming of Tooley's "no future" space-time model might be pointed out. In the EPR experiment, no explanation of the correlations between the outcomes of the measurement events A and B is yet forthcoming. In Tooley's model, when the outcome + or spin-up is added as a new event just later than A, then - is also added as a new event just later than B if the magnets are aligned. But what coordinates or orchestrates these new additions? If the magnets are misaligned and the outcome of A is +, then the probability of getting + at B is p; if the outcome at A had been - instead of +, the probability of getting + at B would have been q. What mechanism lies behind these correlations? How does the electron at B know its probability of passing + following the outcome at A? In the "no future" ontology, any connection between adding events at A and at B is unexplained. If on the other hand the future is not empty, but contains possible future events from which those that become actual are selected, then, as in my *Model* (1994, pp. 94-114), a physical mechanism which accounts for the mysterious non-local connections becomes available.

REFERENCES

Broad, C. D. (1923), *Scientific Thought* (London: Routledge and Kegan Paul Ltd).

Broad, C. D. (1959), "A Reply to My Critics," in P. A. Schlipp (ed.), *The Philosophy of C. D. Broad* (La Salle, Ill: Open Court): 711-830.

McCall, S. (1994), *A Model of the Universe* (Oxford: Clarendon Press).

Tooley, M. (1997), *Time, Tense, and Causation* (Oxford: Oxford University Press).

3
ACTUALITY AND ACTUALITY AS OF A TIME

Quentin Smith
Western Michigan University

Readers of Michael Tooley's excellent book *Causation: A Realist Approach* (1987) will rightly predict that *Time, Tense, and Causation* will also contain many interesting and highly original theses that are supported by some intricate precisely formulated arguments. The main thesis of his new book is that the tensed or dynamic theory of time does not require any irreducible A- properties, tensed concepts, tensed propositions, tensed facts or tensed truth conditions, and implies instead that reality consists only of irreducible tenseless facts, tenseless concepts, tenseless propositions and like. According to Tooley, the thesis that the tensed theory of time is true, that the future is unreal and the present and past are real, and yet that reality consists only of tenseless facts, etc., is neither self-contradictory nor false. He rejects five familiar tenets of the tensed theory of time:

(1) Tenseless temporal concepts are analyzable in terms of tensed temporal concepts, such as the concepts of the past, present and future.
(2) The concepts of past, present and future are semantically basic.
(3) Familiar tensed sentences do not include indexicals and express the same proposition at different times of utterance.
(4) There is no conceptually basic, tenseless existential quantifier that ranges over all entities.
(5) The classical concept of truth should be rejected in favor of a temporally indexed concept of truth.

Tooley's thesis that the tensed theory of time is true but there are only tenseless facts seems to be a contradiction on the face of it, so how does Tooley argue that this theory is consistent? He holds that the future is unreal and the present and past are real, and that this idea is analyzable in terms of the notion of actuality as of time. At a given time, say 1980, the states of affairs that are simultaneous with or earlier than 1980 are actual as of that time, but the states of affairs that are later than 1980 are not actual as of 1980. This immediately brings us to the fundamental idea in Tooley's

L. N. Oaklander (ed.), The Importance of Time, 21–29.

theory, actuality as of a time. Exactly what does it mean to say that something is "actual as of a time?"

Tooley claims that this phrase expresses a primitive, unanalyzable notion. It is precisely this claim that both tensers and detensers will fasten upon as the most questionable element in Tooley's theory. This allegedly primitive notion is dissimilar to the primitives with which philosophers of time are normally familiar. His primitive, unanalyzable notion does not mean, for example, the tenseless notion of *being an event that occurs simultaneously with a time t*, and it does not mean the tensed notion expressed by "*is present*" or "*occurs at whatever time is now.*" Nor does it mean *is simultaneous* or *earlier than a time t* or the irreducible tensed notion of *is present or past*. What, then, could it mean?

Since it is unanalyzable, Tooley cannot explain what it means. But it could be argued that it is unobjectionable to take it as primitive. But does he do this? Tooley says he will argue "in this section [2.2]" (1997, p. 37) that there is nothing objectionable about treating his notion as primitive. However, after rereading the section in question several times (1997, pp. 37-42) I could not find any such argument. He argues that there is nothing objectionable about taking the *different* notion of *actuality simpliciter* as a primitive (1997, pp. 37-39).

He also presents two informally invalid arguments that the advocate of dynamic theory of time needs both the notion of actuality *simpliciter* and the notion of actuality as of time (1997, p. 40). The first argument is that propositions, numbers and uninstantiated universals do not exist in time and therefore that a distinction is needed between *actuality simpliciter* and *actuality as of a time*. This begs the question against Prior, Wolterstorff, myself and a long line of tensers (proponents of a dynamic theory of time) who argue that abstract objects exist in time. Tooley makes no attempt to answer the many arguments offered for the temporal existence of abstract objects.

Second, he says we need the distinction between actuality *simpliciter* and actuality as of a time since "besides being able to speak of what is actual as of the present moment, one needs also to be able to make sense of propositions about what was actual, about what will be actual, and more generally, about the totality of what is actual at some time or other..." (1997, p. 40). But this argument includes in its premises one of the notions Tooley is arguing is necessary to distinguish; the premises assume the notion of what is actual as of a time and proceed from this premise to argue that we need in addition the distinct notion of the totality of what is actual at some time or other. This argument is thus also a petitio principii and, in any case, is not an argument for the thesis that there is nothing objectionable about taking the notion of actuality as of a time as primitive.

Tooley then proceeds to offer another argument, but it is also not an argument that "there is nothing objectionable" about the view that the notion of actually as of a time is primitive. Rather, this next argument is that the notion of *being an actual temporal state of affairs* cannot be analyzed in terms of the notion of being actual as of a time since (on the dynamic view) it is not the case that every time is actual as of every other time. But this argument is not the one promised, since if a concept X cannot be analyzed in terms of a concept Y, it does not follow that Y is unanalyzable. Tooley concludes: "Consequently, it seems that, on any dynamic view of the world, there are two concepts that need to be taken as primitive" (1997, p. 41). However, even if we assume the earlier arguments in this section are valid, all that follows is that dynamic view requires a notion of *actuality* that is *distinct* from the notion of *actuality as of a time*. None of the arguments have as their conclusion or "consequence" that the notion of actuality as of a time is primitive (or that there is nothing objectionable about taking this notion as primitive).

An examination of the argument Tooley gives for the thesis that the different notion of actuality *simpliciter* is primitive reveals that this conclusion also does not follow from his premises. He rejects Adam's definition of "is actual" in terms of the notion of a proposition being true and says "whether something is actual does not, unlike the case of a proposition's being true, seem to be matter of a relationship to something else" (1997, p. 39). However the premise that actuality is not a relationship to something else does not entail that the notion of actuality is primitive, since this notion could be reductively definable in terms of a relationship of something to itself or in terms of a monadic property. For example, the notion of actuality could be reductively definable in terms of the primitive notion self-identical (a relationship of something to itself), or in terms of the monadic property of being physical (giving a materialist ontology) or being a sense datum (a phenomenalism that represents the universe as a collection of sense data). These reductive analyses of actuality *simpliciter* may be false, but Tooley gives no argument against them or against other such reductive analyses. Since Tooley's entire theory rests on the claim that there is a distinction between the primitive notion of actuality and the primitive notion of actuality as of a time, and since either no argument or no valid argument is given that there are such notions, the foundation of his theory lacks the firmness that one might desire.

It might seem that this result shows that Tooley's theory lacks the clear foundations that enables his arguments to be followed. But I suggest his theory can be reinterpreted in ways that makes his arguments clear enough to be addressed. I suggest that an examination of the role that the phrase "actual as of a time t" plays in Tooley's book shows that he mostly uses this phrase to express the same concept as "exists simultaneously with the time t or earlier that the time t," where "exists" is used tenselessly. This is a tenseless concept and detensers would certainly agree that for each time t, there are some events that are simultaneous with or earlier than t (except for a possible first instant). This reinterpretation would confirm the suspicion that Tooley's initial rejection of the five theses central to the tensed theory of time have the result that his theory is in fact a version of the tenseless theory of time, despite Tooley's protestations to the contrary.

But this is not entirely fair to Tooley since, among other reasons, he adopts the idea that some propositions are neither true or false at some time (say 1998) but are true (say) in 1999. Perhaps this is where the "dynamic" or tensed theory of time aspect of his theory is located, since this is alleged to capture the notion that the future is unreal but the present and past real. Tooley says it is not true in the year 1980 that there are (tenselessly) unicorns in the year AD 2000. But it is also not false in the year 1980 that there are (tenselessly) unicorns in the year AD 2000. Rather, it has a third truth value in 1980, indeterminateness, since "there is nothing actual, as of that time, that makes [it] true" (1997, p. 129) and there is no positive truthmaker for some logically incompatible statement that is actual as of 1980. Unfortunately, the distinction between "true" and "true at a time" is based on the distinction between the primitive notions of actuality *simpliciter* and actuality as of a time, so we return to the original problems we have with his arguments about these two notions of actuality.

But here again his theory can be clearly represented if we choose to interpret it in terms of the tenseless theory of time. In fact, the way Tooley uses the phrases involving "truth" suggest that he has in mind a notion that belongs to the tenseless theory of time. I think defenders of the tenseless theory of time will grant that "there are (tenseless) 27 hurricanes in the year AD 2000" has the truth value of indeterminateness in the year 1980 if this is merely a imprecise or cumbersome way of saying that "there are no relevant truthmakers of false-makers that are simultaneous with or earlier than 1980." In other words, the proposition is indeterminate relative to the truthmakers and false-makers that exist in 1980 and earlier.

But if Tooley adds that it also implies that there are no relevant truthmakers or false-makers later than 1980, he is adding something inconsistent with his notion of truth *simpliciter*, for he holds it is true *simpliciter* that (for example) there are 27 hurricanes in the year AD 2000. He says "the future is part of the totality of reality and so is actual *simpliciter*" (1997, p. 303). He cannot consistently say that the proposition about hurricanes in 2000 is indeterminate relative to the truth and false-makers that are later than 1980, since included in these truthmakers is the occurrence of 27 hurricanes in the year 2000 and thus relative to these truthmakers the proposition is true. This cannot be denied since Tooley holds that the hurricanes in 2000 are actual *simpliciter*.

The notion of "truth at a time" proves to carry no ontological weight, since one could equally well introduce a phrase such as "truth pertinent to a time" and say that a proposition is true pertinent to 1980 if and only if its truth or false-makers are simultaneous with or later than 1980. There are no concepts here that are not a part of the standard tenseless theory of time. We have introduced new stipulated uses of certain phrases, but theses do not introduce concepts that are not part of the standard tenseless theory. There are no indeterminate truth values, since the alleged indeterminateness is obtained by conceptually or mentally abstracting from the part of actuality in which the relevant truth and false makers exist. The proposition is true, but I can make it appear to me be indeterminate by paying attention only to the times earlier than the times at which its truthmaker exists.

Tooley might nonetheless object that his theory is distinct from the tenseless theory of time. However, since Tooley disallows the irreducible tensed facts that 1980 is present and that 2000 is future, there is no way in his ontology to distinguish in ontological status the year 2000 from the year 1980. For any time t, be it 1980, 1998 or 2000, the "future is unactual as of that time" in the sense that the future is later than that time.

But Tooley may have a counter to this argument in his book. Tooley's apparent counter is his argument (e.g., on page 211) that sentences such as "In the year 1900, the birth of Bertrand Russell is an event that lies in the past" is not equivalent in meaning to "The birth of Bertrand Russell is earlier than the year 1900" (as defenders of the tenseless theory of time would say), since the first sentence alone entails that "what is actual as of the year 1900 does not include any states of affairs that are later than 1900." I demur, since both sentences entail a sentence equivalent to Tooley's sentence about what is actual as of 1900, the equivalent sentence: "what is (wholly) simultaneous with or earlier than 1900 does not include any state of affairs that is later than 1900." This again is part of the standard tenseless theory of time.

I suggest that what Tooley claims to be a primitive notion of actuality as of a time is really the familiar notion in the tenseless theory of time that is expressed by "being simultaneous with or earlier than a time," taken in conjunction with a decision to mentally or conceptually abstract from what is later than that time. Tooley's theory implies that for each time t, there is something later than that time (except for a possible last instant of time). Tooley says "the future is part of the totality of reality, and so is actual *simpliciter*" (1997, p. 303). When Tooley goes on to say that "the future is not actual as of the present moment" (1997, p. 303) one may respond that, given Tooley's rejection of tensed facts, this sentence (if uttered at March 1,

1998) is equivalent to "what is later than March 1, 1998 is neither simultaneous with nor earlier than March 1, 1998." Tooley would respond that this equivalence does not obtain since it fails to capture the primitive, unanalyzable notion of being actual as of March 1, 1998. Why cannot the tenser and detenser respond that there is no such understandable notion?

Further, tensers and detensers may even respond that Tooley's appeal to such a notion involves an implicit contradiction, namely, that his corresponding distinction between truth *simpliciter* and truth at a time allows for such inferences as the following one. Item (1)

(1) The sentence "There are at least two hurricanes in 2002 A.D." is true

Implies

(2) Some 1998 token of "at least two hurricanes will occur four years from now" is not true in 1998.

This is a contradiction, since (1) and (2) cannot both be true. Suppose sentence (1) is true. Sentence (1) implies that four years later than 1998, at least two hurricanes occur. This in turn implies that if it is now 1998, then four years later at least two hurricanes will occur. This in turn implies that if the sentence "at least two hurricanes will occur four years from now" is uttered 1998, so that "now" refers to 1998, then it is true. Since Tooley theory implies that if (1) is true, (2) is also true, his theory implies a contradiction.

If Tooley responds that there are no contradictions since there are primitive, unanalyzable notions of actuality *simpliciter* and actuality as of a time that guarantees that there are no contradictions of the sort that I have pointed out, then we are back again to the counterresponse of the standard tensers and detensers that they can understand no such primitive notions.

Tooley's theory would be advantaged if he could successfully argue that the five standard theses of the tensed theory of time are false. But his arguments in this regard may also be questioned. For example, Tooley argues that it is a contradiction to suppose that instantaneous events have intrinsic or monadic properties of presentness of pastness. He says it is impossible for an instantaneous event to have different properties at different times.

But why is it impossible for an event e to have the property of presentness at one time and have the property of pastness at later time? Because Tooley stipulates that there are only two possible ways for what is true of an entity to be different from one time to another. "The first is that whether something is true of an object may depend upon how other parts of the world are, and these may be different at different times" (1997, p. 227). But this cannot explain monadic or intrinsic properties of presentness and pastness, Tooley notes, since the rest of the world had no bearing on whether an event possess intrinsic properties. The second possible way is that "the object itself may have temporal parts, with different properties at different times" (1997, p. 227). Tooley says that instantaneous events have no temporal parts and thus that they cannot have different intrinsic properties in this way.

But why should a defender of intrinsic properties of presentness and pastness accept Tooley's stipulation that these are the only two possible ways for something to possess different intrinsic properties at different times? Why can't the defender hold that there is a third way, that with the passage of time some intrinsic temporal properties change, specifically, that when time passes an object can change from possessing the intrinsic property of presentness to possessing the intrinsic property of pastness? As far as I can see, Tooley just stipulates that events cannot have an intrinsic property of presentness at one time and an intrinsic property of pastness at a later time. By contrast, the tenser can not merely counter-stipulate with equal success that events have such properties, but present several dozen arguments for this thesis, arguments that appear in the published literature but are not fully addressed in Tooley's book.

Tooley claims that ordinary tensed sentences are indexical and have truth conditions that involve mention of times, such as time t or specifically, times such as March 1, 1998. Tooley allows that the thesis that tensed sentences have token-reflexive truth conditions was successfully argued in Chapter 3 of my *Language and Time* to "generate the wrong truth-values in certain cases." But Tooley goes on to say that "there is nothing in this objection" (1997, p. 246) since the detenser could supply acceptable truth conditions of the form "E is simultaneous with t" or (as on Tooley's own account) of the form "E lies in the present at time t." This assertion reveals a problem in Tooley's theory along the lines of someone jumping from the frying pan into the fire. In Chapter II of *Language and Time*, I provided several different arguments for the thesis it is logically incoherent for ordinary tensed sentences to have truth conditions that mention a time or date, such as t or March 1, 1998. It is not clear why Tooley makes no attempt to respond to such arguments against the date-analyses he uses.

Tooley's argument that an ordinary tensed sentence such as "It is raining in London" express different propositions at different times of utterance is invalid. He says "... a sentence such as 'It is raining in London' contains no expression that specifies the time in question. The relevant time is simply the time at which the sentence is uttered, and, as a consequence, the proposition expressed by an utterance of such sentence a week ago necessarily differs from the proposition expressed by an utterance of it today" (1997, p. 190). If the relevant time is not specified by the sentence but is the time of utterance, it does not follow (contra Tooley's claim) that the proposition expressed a week ago differs from the proposition expressed today. It is consistent to hold that this sentence always expresses the same proposition, namely, that some event of there being rain in London has the property of presentness, and yet that the rules of use of this sentence imply that the time that is ascribed presentness by this proposition is whatever time is the time at which the proposition is expressed by the sentence-utterance.

In fact, however, Tooley's unargued for premise that the tensed sentence must specify a time is false, for it is sufficient for the sentence to be true that the event referred to is specified, and this event is specified by the rules of use of the sentence to be the event that is present when the sentence-utterance is present. Indeed, for the reasons indicated in Chapter 2 of my *Language and Time*, the sentence-utterance cannot specify the time.

Tooley, says that my book *Language and Time* is a "conceptual analysis" of our conceptual framework and that this "provides no grounds for concluding that the world in fact conforms to that framework" (1997, p. 248). Tooley provides no support for this assertion and does not address my contrary argument in Chapter 1.4 that the metaphysical investigations I undertake do provide grounds for concluding that the world in fact conforms to the notion analyzed. I argue there that the sort of skeptical claim Tooley makes about my method of metaphysical argument is self-referentially unjustifiable. When Tooley mentions my theory under the heading of "absence of metaphysical argument" in his index, it does not appear that this is an accurate characterization.

Arguably, it is Tooley's theory that has the problem of being merely an analysis of humans' conceptual frame that provides no grounds for concluding that this analysis can reveal something about the nature of mind-independent reality. For I hold that the propositions and concepts I analyze are mind-independent abstract objects that correspond to mind-independent states of affairs. By contrast, Tooley says the concepts and propositions he is analyzing are mind-dependent "constituents of mental states" (1997, p. 128). Where is Tooley's argument that an analysis of the human species' mental states can show us something about the nature of mind-independent reality?

In this paper I have concentrated on some of the most debatable parts of Tooley's book. But to avoid giving the wrong impression, some of the many positive aspects of the book also deserve mention. For example, he gives an extensive and well argued defense of three-valued logic (truth, falsity, and inderminateness) and of the thesis that disjunction is not truth-functional. He also provides a plausible and careful argument that some spacetime points stand in causal relations to one another. Other examples of good arguments abound.

But perhaps the best part of Tooley's book is the carefully argued chapter on special relativity, where Tooley uses the EPR correlations to motivate the construction of a theory of absolute simultaneity that is as near to STR as possible and that does not require the "non-scientific" assumption that abstract objects exist in time. There has been much handwaving since Popper's initial comment that Aspect's 1981 experiment vindicates a neo-Lorentzian rather than an Einsteinian interpertation of the Lorentz transformations, and Tooley seems to be the first philosopher since the Aspect experiment to work out in detail such a theory based solely on physical principles. (Craig has a good neo-Lorentzian theory in a forthcoming book, but it requires a deity, and my theory of absolute simultaneity requires an appeal to abstract objects). Tooley's does not discuss any of the neo-Lorentz literature (specifically, the pre-1980 work by Prokhonvik, Builder, Ives or the neo-Lorentz essay by J. S. Bell on "How To Teach Special Relativity" etc.) and one might ask if he should have ascribed to the standard neo-Lorentzian thesis that motion through absolute space (or some medium in absolute space) is the cause of time dilation and length contraction. In Tooley's theory it is an inexplicable fact that time dilates and length contracts when bodies move in absolute space. Tooley's neo-Lorentz theory is nonetheless of very high quality and should be required reading for any philosopher or physicist who wants to understand time.

The strengths of Tooley's book considerably outweigh its weaknesses; it is an intricate and thought-provoking challenge to defenders of standard tensed or tenseless theories of time and in this respect as well as many others (e.g., his good discussions of causation, counterfactuals, three-valued logic, disjunction, STR, etc.) it is an important contribution to the field and will be a book that philosophers of time will be discussing quite extensively.

4
RESPONSE TO THE COMMENTS ON *TIME, TENSE, AND CAUSATION* BY STORRS MCCALL, NATHAN OAKLANDER, AND QUENTIN SMITH

Michael Tooley
University of Colorado at Boulder

1. RESPONSE TO STORRS MCCALL'S COMMENTS

Let me begin with the gentlest of my three critics - Professor Storrs McCall. With regard to some very fundamental issues in the philosophy of time, his views are in some ways, as we shall see below, close to my own. At the same time, there are some very important differences. Thus, on the one hand, I hold a view of time that, as Storrs McCall points out, is quite similar in some respects to the view advanced by C. D. Broad in his book, *Scientific Thought* (1923, pp. 53-84) - since it involves, first of all, the idea that while the past and the present are now real, the future is not, and secondly, the idea that the world grows over time by the accretion of new facts or states of affairs. Storrs McCall, by contrast, has put forward a rather different view of time in his very interesting and important book, *A Model of the Universe* (1994). On that view, the world, rather than growing by the accretion of new facts, shrinks, so to speak, by the deletion of facts. But McCall is not putting forward what might be called a "flipped-over" Broad-type view - to the effect that while the future and the present are real, the past is not. (It is risky, in the philosophy of time, to claim that no one has ever held a certain view. But in the case of the flipped-over Broad-type view, a list of advocates does not, at least, readily spring to mind.) McCall's view is rather that what gets deleted with the passage of time is simply the future possibilities that are not actualized - possibilities that McCall views in a fashion similar to the way that David Lewis views other possible worlds - namely, as things that are no less concrete than past facts or present facts.

L. N. Oaklander (ed.), The Importance of Time, 31–58.

This is certainly an important difference, and McCall and I both think that there are distinct advantages to the model that each of us, respectively, favors. Thus, as he mentions in his comments, he feels, for example, that his model for a dynamic world enables him better to account for certain quantum mechanical correlations that one finds in Einstein-Podolsky-Rosen-type situations. I feel, on the other hand, that a Broad-type model is preferable because, among other things, it avoids extreme realism concerning future possibilities, and because it makes it possible for even a completely deterministic world to be a dynamic world.

These differences notwithstanding, McCall and I are, I believe, really quite close on a number of fundamental issues concerning the ontology of time, since, in addition to sharing a dynamic view of the world, we also share the idea that the history of the world consists of tenseless states of affairs, with tensed facts being logically supervenient upon what tenseless facts are actual as of different times.

McCall devoted the first part of his paper to arguing that there is a satisfactory answer to the basic objection raised by Professor Oaklander - and also, in effect, to the objection advanced by Quentin Smith in the first part of his paper. I found that discussion very clear and very elegant, and - it seemed to me - totally convincing. I hasten to add, of course, that the other two members of our panel may perhaps not have been quite as convinced!

More generally, I found relatively little to disagree with in Storrs McCall's discussion as a whole. But there were two places - which seem to be of some importance - where I was not convinced. The first concerns the idea of truth at a time; the second, my appeal to quantum mechanics in support of the idea of absolute simultaneity.

1.1 Is Truth at a Time an Optional Extra?

As regards the first of these issues, what needs to be kept in focus - and it will probably not be hard to do so, given that more than one of the commentators has emphasized it - is that I hold, in contrast to all advocates of static approaches to time, and the vast majority of advocates of dynamic approaches to time - that both the concept of what is actual *simpliciter*, and the concept of what is actual as of a time, are crucial for a sound account of the nature of time. McCall, as he has indicated, thinks that this position is defensible. In addition, however, I also embrace both the idea of truth *simpliciter* and the idea of truth at a time, and McCall raises the question of whether one should go down that road. He suggests that I might have refrained from doing so by agreeing with Ducasse, who held that to qualify the terms "truth" and "falsity" by a date is just as problematic as to qualify those terms by a place, and so I could, McCall suggests, hold that truth at a time does not make sense, and that the only legitimate concept of truth is that of truth *simpliciter*.

In response, I do not want to make any claims about the concept of truth in natural languages: it may well be that a temporally-indexed concept of truth is not to be found in such languages. But even if that is so, can one avoid the claim, given the metaphysics I embrace, that the concept of truth at a time must be definable, and intelligible? I do not see that one can. For if truth *simpliciter* is a matter of correspondence between a proposition, or statement, or sentence (or other bearer of truth-values) and what is actual *simpliciter*, then must it not be possible, employing the same correspondence relation, to introduce a temporally-indexed concept of truth that is a matter of a similar relation of correspondence between propositions (or other bearers of truth-values) and what is actual as of a time?

In short, given the ontological distinction between what is actual *simpliciter* and what is actual as of a time, can one rule out a corresponding semantical distinction between truth *simpliciter* and truth at a time? I do not see how one can, and so it seems to me that once one accepts the idea of being actual as of a time, one must grant that the concept of truth at a time can be defined.

1.2 Quantum Mechanics and Absolute Simultaneity

One of the objections that any dynamic view of time must grapple with is based upon the Special Theory of Relativity, and the thrust of it is that, on the one hand, dynamic conceptions of time typically involve the idea of a universe-wide present, and thus the idea of absolute simultaneity, whereas, on the other hand, the Special Theory of Relativity does not involve any idea of a relation of absolute simultaneity that holds between events. So dynamic theories of the nature of time appear, in general, to be on a collision course with a very well-confirmed scientific theory.

One possible response to this problem is to modify such metaphysical theories – perhaps along the lines proposed by Howard Stein in his paper, "On Einstein-Minkowski Space-Time" (1968) – and to abandon the idea of events being actual *as of a time*, replacing it with the idea of events being actual *as of a spacetime point*. If possible, however, I would like to avoid going along that route, and so I attempted, instead, to offer arguments in favor of absolute simultaneity.

The main argument that I offered involved the idea that it is reasonable to believe that there are causal relations between spacetime points, and then, given that idea, I asked what hypothesis concerning the details of that relationship is most plausible. In response, I argued that it was more likely that the causal connections were non-branching, rather than branching, and on this basis I was able to develop an argument for absolute simultaneity. (The same line of thought supports the idea of absolute location within Newtonian physics.)

Such was, in outline, my main argument. In addition, however - and ignoring the advice of a good friend, who holds that an appeal to quantum mechanics is the last refuge of a philosophical scoundrel - I also set out an argument based upon quantum mechanics. (But while ignoring the advice of my friend, I was at least following the advice of one of my high school teachers, who liked to say: "Do not pass temptation lightly by. It may never come again!") So, succumbing to temptation, I appealed to quantum mechanics, and argued that if one considered, for example, a pair of electrons created in an entangled spin state, then a measurement on either would collapse the wave packet, so that there would be a time at which the electron on which no measurement had been made would also have acquired a determinate spin. The times at which the two electrons acquired determinate spins would thus determine, accordingly, a relation of absolute simultaneity.

McCall contends that this line of argument is not sound, his objection being that according to the traditional, Copenhagen interpretation of quantum mechanics, "definite values of physical quantities such as spin cannot be assigned to quantum systems unless a 'measurement' of that quantity is made." So the idea is that, even after a measurement has been made of the spin of the first electron, the second electron does not really have a spin until a measurement is made *on it*.

The problem that I have with this argument is that, even if one grants, for the sake of discussion, that the second electron has no spin until a measurement is made on it, it would seem to be true that what spin it would have, if it were to be measured, is determinate, and so, once a measurement has been made on the first electron, the *probability* that the second electron will have just the spin that is required by the relevant conservation principle must be, *at some point, equal to one* - whereas before any measurement was made on the first electron, the probability in question was not equal to one But then one can ask at *what* point the probability changed from not being equal to one to being equal to one, and it would seem that the answer to that question will once again determine a relation of absolute simultaneity. So, although the philosophy of quantum physics is an area where Storrs McCall is much more knowledgeable than I am, and thus an area where I do not have enormous confidence in my judgments, it still seems to me that both the general idea of the collapse of the wave packet, and the specific case of EPR-type situations, provide grounds for thinking that the world must contain events that stand in relations of absolute simultaneity.

2. RESPONSE TO NATHAN OAKLANDER'S COMMENTS

2.1 Some Background Concerning my Views on Time

As Nathan Oaklander emphasizes, the dynamic conception of the world that I am defending differs from most tensed accounts of time in a number of important respects. Of these differences, perhaps the most important is that, rather than holding that tenseless facts are logically supervenient upon tensed facts, I hold that tensed facts are logically supervenient upon what tenseless facts are actual as of different times.

Another important difference involves, first, the ontological concept of being actual, and the semantical concept of truth. As regards the former, and in common with all defenders of tensed or dynamic views of time, I must hold that there is a coherent, and non-trivial, temporally-indexed notion of actuality, so that what is actual as of one time may - and indeed must - differ from what is actual as of any other time. Different advocates of dynamic views of time will, admittedly, have different opinions concerning the most basic way of referring to different times. Most hold that this should be done by means of tensed expressions - and so they will talk about what is actual now in contrast to what was actual, say, five minutes in the past. I hold, by contrast, that one can refer to moments of time either by tenseless expressions or by tensed expressions, and that, contrary to what is claimed by most advocates of tensed approaches to time, it is not the case that reference by means of tenseless expressions must, or even can, be analyzed in tensed terms, since in order to do that one would have to be able to analyze temporal priority in tensed terms, and the latter cannot be done (Tooley (1997), pp. 153-173). But these disagreements about the relations between tensed and tenseless temporal expressions do not affect the basic point here - namely, that dynamic accounts of the nature of time require a temporally-indexed notion of actuality.

This temporally-indexed ontological concept generates, in turn - as I have just argued in my response to McCall - a temporally-indexed concept of truth, according to which a proposition may have different truth values at different times. (In contrast to some other tensed views, I maintain, however, that a proposition can change its truth value only once, and that that change must be from having the third truth value - indeterminateness - to being true, or to being false.)

Where my approach diverges from that of other tensed accounts is that I hold that, in addition to the temporally-indexed notions of what is actual as of a time, and of truth at a time, one also needs the absolute notions of what is actual *simpliciter*, and of truth *simpliciter*. Why so? One important reason is briefly mentioned by Oaklander, and it concerns the concept of the future. Contrary to other advocates of tensed views, I maintain that the concept of the future cannot be taken as analytically basic, on the grounds that semantically basic concepts (of a non-logical sort) must apply in virtue of the presence of properties or relations with which one can be directly acquainted, and that, since we are not directly acquainted with future events, it cannot be the case that we are directly acquainted with the property of futurity. So the concept of the future stands in need of *analysis*, and when the question of how this can be done is pursued, it becomes clear that one needs quantifiers that range over everything that exists at any time, past, present, and future. But one cannot make sense of such quantifiers unless one can make sense of the concept of what is actual *simpliciter*.

Given the concept of what is actual *simpliciter*, it is possible to define the notion of truth *simpliciter*. But I argue that there are also other reasons for accepting the notion of truth *simpliciter*. For one thing - and here I am disagreeing with a view advanced by Quentin Smith - I do not think that it is plausible to employ the concept of truth at a time in the case of statements - such as mathematical statements - that do not refer to temporal entities. For another thing, if one considers quantified statements, it is clear that there can be statements that, although they are never true *at* any time, may nevertheless be true. (Consider, for example: "Every temporal moment is always followed by another temporal moment.")

How is what is actual as of a given time, *t*, related to what is actual *simpliciter*? The relationship is a simple one, on my view, due to the fact that I hold that the world changes only by the addition of tenseless facts. This view contrasts, of course, with typical tensed views, which take tensed facts as basic, and which thus are forced to say that the passage of time involves the addition of new states of affairs, accompanied by the deletion of at least some tensed states of affairs that previously existed, since every instantaneous event is present only for an instant. Moreover, given such views, the notion of states of affairs that are actual *simpliciter* seems problematic. But if, as I maintain, the world changes only by the addition of tenseless facts, then if t_1 and t_2 are any two times such that t_2 is later than t_1, then what is actual as of time t_1 is a proper part of what is actual as of t_2. This means that one can form the mereological whole of everything that is actual at some time or other, and one will have a consistent state of affairs, and one that is identical with the totality of what is actual *simpliciter*.

2.2 Response to Oaklander's First Objection

One upshot is that, on my approach, the following two propositions are true:

(1) What is actual *simpliciter* is identical with the totality of everything that exists tenselessly;
(2) What is actual *simpliciter* is identical with the totality of everything that is actual as of some time or other.

Oaklander suggests that in embracing both the idea of what is actual as of a time, and the idea of what is actual *simpliciter*, I am trying to have my cake and eat it too. What does he see as the problem here? His answer is given in the following, crucial passage:

> Tooley never clearly explains how the sum total of existence can remain the same and have temporal parts that change their existential status. Tooley asserts that the problem of change motivates the debate between tensed and tenseless approaches to time (1997, p. 13), but he does not realize that his own solution simply reraises it. He maintains that the totality exists *simpliciter* and yet new tenseless facts come into existence and are added on to the totality as time passes. But how is this possible? How can the totality exist *simpliciter* and remain the same through a change (as of different times) in the existential status of its temporal parts. (p. 6)

Oaklander has formulated his criticism in terms of a problem, and has asked whether the problem can be solved. It may help, however, to recast it in a slightly less gentle form, as an argument:

(1) The totality of existence is always the same.
(2) Every contingent state of affairs changes its existential status, since it comes into existence at some point in time.
(3) The totality of existence has, as parts, absolutely all contingent states of affairs.
(4) Therefore, the totality of existence has parts that come into existence. [From (2) and (3)]

(5) If some totality has a part that comes into existence, then that totality cannot always be the same.
(6) Therefore, the totality of existence cannot always be the same. [From (4) and (5)]
(7) Statements (1) and (6) are logically incompatible.
(8) So statements (1), (2), (3), and (5) cannot all be true, since they lead to a contradiction.

My response to this is that if one accepts both the idea of being actual as of a time, and of being actual *simpliciter*, then one cannot speak simply of "*the* totality." One has to speak, on the one hand, of *the totality of what is actual simpliciter*, and of *the totality of what is actual as of any given time t.* Consider, now, Oaklander's question: "How can the totality exist *simpliciter* and remain the same through a change (as of different times) in the existential status of its temporal parts?" The answer to the first part of this question – "How can the totality of what is actual *simpliciter* remain the same?" – is just the answer that would be given by an advocate of a tenseless view of time if asked how the totality of all facts can remain the same – namely, that *that* totality is not temporally indexed, and so it is logically impossible for it to change. The second part of Oaklander's question, however, is suggesting that I cannot give the same answer that an advocate of a tenseless view would give, since I hold that parts of that totality change their "existential status." But the expression "existential status" obscures the distinction I have just mentioned – between the totality of what is actual *simpliciter*, and the totality of what is actual as of any given time *t*. For what the change in "existential status" comes down to is that the totality of what is actual as of any one time is necessarily different from the totality of what is actual as of any other time. But this is not to say that what is actual *simpliciter* is different at one time from what is actual *simpliciter* at any other time. My concept of the facts that are actual *simpliciter* is the very same concept as that employed by advocates of tenseless views of time, and it is incoherent to combine *that* concept with any temporal indexing.

Consider, now, how the argument just set out reads when the distinction between being actual *simpliciter* and being actual as of a time is introduced. In the case of statements (1) through (4), it is clear, I think, how they should be formulated:

(1) The totality of what is actual *simpliciter* cannot be different at different times, since the concept of what is actual *simpliciter* is not a temporally-indexed concept, and so it makes no sense to say that what is actual *simpliciter* is different at different times.

(2) Every contingent state of affairs changes its existential status, since it is actual as of some times, and not actual as of others.

(3) The totality of what is actual *simpliciter* has, as parts, all contingent states of affairs.

(4) Therefore, the totality of what is actual *simpliciter* has parts that are actual as of some times, and not actual as of others. [From (2) and (3)]

So formulated, statements (1) through (4) are true, and the inference from statements (2) and (3) to statement (4) is valid. But how is statement (5) to be formulated? Two different possibilities suggest themselves:

(5a) If some totality has a part that is actual as of some times, and not actual as of others, then that totality cannot be actual as of every time.

(5b) If some totality has a part that is actual as of some times, and not actual as of others, then that totality cannot be actual *simpliciter*.

If one goes with interpretation (5a), it together with statement (4) will entail:

(6a) Therefore, the totality of existence cannot be actual as of every time. [From (4) and (5a)]

On the other hand, if one goes with interpretation (5b), it together with statement (4) will entail:

(6b) Therefore, the totality of existence cannot be actual *simpliciter*. [From (4) and (5b)]

How does the argument fare under these two readings? In the case of the first reading, statement (5a) is indeed true, but this is not troubling since (6a) is not inconsistent with statement (1), and so one does not have a contradiction: to say that some state of affairs – be it a totality or a non-totality – is actual *simpliciter* does not entail that it is actual as of *every* time. In the case of the second reading, on the other hand, one does have a contradiction: statement (6b) is inconsistent with statement (1). But here the argument fails because statement (5b) is false, as is clear once one notices that anything that is actual as of at least one time is *ipso facto* – given the view that the past is real – actual *simpliciter*, and since every part of the totality is actual *simpliciter*, there is no problem with the totality's being actual *simpliciter*. Conclusion: once the distinction between being actual as of a time and being actual *simpliciter* is introduced, the argument that is suggested by Oaklander's challenge collapses.

What I am saying, in short, is this:

(1) There is a coherent, temporally-indexed notion of being actual, and, given that notion, what is actual as of one time necessarily differs from what is actual as of any other time.

(2) There is also a coherent, absolute notion of being actual *simpliciter*, and, given that notion, it is incoherent to talk about what is actual *simpliciter* as of a time, let alone to say that what is actual *simpliciter* as of one time may differ from what is actual *simpliciter* as of some other time.

(3) Any state of affairs that is actual as of some time *t* is actual *simpliciter*.

(4) Claims (1), (2), and (3) are mutually consistent.

2.3 Where Do Oaklander's Doubts Really Lie?

I suspect that what really lies behind Oaklander's question, however, is not the idea that a contradiction arises if one accepts *both* the concept of being actual *simpliciter and* the concept of being actual as of a time, but the familiar tenseless view that the concept of being actual as of a time is *itself*, in the end, somehow unintelligible, so that the whole idea of a dynamic world makes no sense. This emerges when Oaklander goes on to consider my account of the meaning of *non-indexical* tensed sentences, which, in the case of the sentence "Event *E* is future at time *t*," is as follows:

> Event *E* is later than time *t*, and *t* is an instantaneous state of affairs, *t* is actual as of time *t*, and no state of affairs that is later than *t* is actual as of time *t*.

Oaklander's comment on this involves asking how it can be the case that " 'no state of affairs that is later than *t* can be actual as of time *t*' unless Tooley simply means that no state of affairs later than *t* is located at *t*" (pp. 6-7). But this is just to challenge the whole idea of being actual as of a time, and with it, the very idea of a dynamic world. So it seems to me that what appears initially to be an objection to the fact that my approach involves acceptance of both the concept of being actual *simpliciter* and the concept of being actual as of a time is, in the end, simply a general rejection of the idea of a dynamic world.

Further confirmation of this is provided, I suggest, by Oaklander's comments on my contention that McTaggart's argument does not have any force at all against the view of time that I advance. (My response to McTaggart is that, since tenseless temporal relations cannot be defined in tensed terms, there can be no objection to using *dates* to specify when a given event has different tensed properties.) Oaklander objects that in holding that statements about the future, though not true now, are true *simpliciter*, a tenseless state of affairs such as Bill Clinton's leaving office in the year 2001 both "exists *simpliciter* and also comes into existence at a certain time," and so "one and the same state of affairs both exists and does not exist" (p. 8). He then says:

> To avoid the contradiction by asserting that these are two different primitive notions of existence strikes me as obfuscating. (p. 8)

Oaklander appears to be saying, then, that it cannot be the case that both the concept of being actual *simpliciter* and the concept of being actual as of a time are primitive notions. But I think that he probably wants to advance a slightly stronger claim here, namely:

> Either the concept of being actual *simpliciter* must be analyzable in terms of the concept of being actual as of a time (possibly together with some other notion or notions), or else the concept of being actual as of a time must be analyzable in terms of the concept of being actual *simpliciter* (possibly together with some other *static* notion or notions).

If he is advancing this claim, then he is advancing a claim that entails that *no* dynamic view of the world can possibly be correct. The reason is as follows:

(1) The concept of being actual *simpliciter* cannot be analyzed in terms of the notion of being actual as of a time, or in terms of other dynamic notions. It must either be taken as primitive, or else analyzed in terms of static notions, such as that of the tenseless existential quantifier.

(2) So the concept of being actual as of a time, if it is intelligible, must be analyzable in terms of the concept of being actual *simpliciter* (possibly together with some other *static* notion or notions).

(3) A dynamic world is by definition a world where what is actual as of one time is different from what is actual as of another time.

(4) If one analyzes the concept of being actual as of a time in tenseless terms - for example, if one says, as Quentin Smith suggests, that the concept of being actual as of time *t* can be defined as being either simultaneous with *t*, or earlier than *t* - then one will not be able to use the concept of being actual as of time *t* to capture the idea of a dynamic world.

(5) Therefore, the idea of a dynamic world is, it turns out, unintelligible.

2.4 Actuality as of a Time, Actuality Simpliciter, and the Existence of a Totality

Oaklander points out that on my approach to time there is such a thing as the totality of reality that is actual *simpliciter*, and he connects this up with my acceptance of the concept of a totality of states of affairs that are actual *simpliciter*. It is important to note, however, first, that the underlying idea here is not really concerned with totalities; it is, instead, the idea that what is actual as of any particular time must also be actual *simpliciter*. Secondly, the reason that I am able to hold that this is the case is that I reject the view that there are intrinsic tensed properties. For if such properties existed, the idea that whatever is actual as of any time must be actual *simpliciter* would mean that what was actual *simpliciter* involved inconsistent states of affairs.

2.5 Further Objections

Later, Oaklander again argues that there is something incoherent in my employing both the idea of being actual as of a time, and the idea of being actual *simpliciter*:

> But is not the very concept of a "total, dynamic world" a contradiction in terms? How can the world be a totality of existence and be dynamic? Is it not absurd to suppose that in addition to tenselessly existing at the time they do, there is some further change that tenseless states of affairs undergo "when" they become actual or come into existence? Indeed, what sense can be made of the idea that new *tenseless* facts get added to what exists *simpliciter*? If the totality exists *simpliciter* then there is (tenselessly) nothing that can be added to them. (p. 10)

There seem to me to be at least three different objections here. First, there is the claim that the idea of a total dynamic world is a contradiction in terms. But why should that be so? To make sense of the idea of a dynamic world, one needs the idea of states of affairs that are actual as of a time. But suppose that the time in question is some time later than the present – such as January 1, 2010. Can one make sense of the idea of states of affairs that are actual as of January 1, 2010? To argue that someone who accepts the view that the world is dynamic cannot make sense of the idea of states of affairs that are actual as of January 1, 2010, it would seem that one must argue that such a person can make no sense of the idea of times that are later than the present, and I see no reason to accept that claim. But suppose, then, that one can make sense of the idea of states of affairs that are actual as of January 1, 2010. Surely one can also make sense, in that case, of the idea of states of affairs such that there is *some* time t such that the states of affairs are actual as of time t. Can one then form the mereological union of all such states of affairs? That will depend, as noted earlier, upon whether the world contains intrinsic tensed properties. If it does, there will be states of affairs that are actual as of different times that are not mutually compatible. But if there are no intrinsic tensed properties, then states of affairs that are actual as of one time will be consistent with states of affairs that are actual as of any other time, and so one will have a consistent whole that consists of all and only states of affairs that are actual as of one time or another.

Secondly, Oaklander asks whether it isn't "absurd to suppose that in addition to tenselessly existing at the time they do, there is some further change that tenseless states of affairs undergo "when" they become actual or come into existence." Here he seems to be asking whether it makes sense to say not only that a certain state of affairs, S, occurs at time t – in the sense of being part of what is simultaneous with t - but also that S is actual as of time t, but not actual as of any time earlier than t. The answer is that this makes sense if and only if the idea of being actual as of a time makes sense. So here, too, Oaklander seems simply to be saying that the idea that is essential to the concept of a dynamic world - the idea of being actual as of a time - makes no sense.

Oaklander's third objection in the above passage is that it makes no sense to think of tenseless facts being added to what exists *simpliciter*: "If the totality exists *simpliciter* then there is (tenselessly) nothing that can be added to them." Here the answer is that Oaklander is quite right that it makes no sense to speak of what is actual *simpliciter* being different at different times. But that is not the view that is being advanced. What is being claimed, rather, is that if time t_1 is earlier than time t_2, then what is actual as of time t_2 includes every tenseless state of affairs that is actual as of time t_1, and more. States of affairs are therefore added to what is actual as of earlier times to give one what is actual as of later times.

A bit later, Oaklander focuses on the combination of the following two claims: (1) The future is part of the totality of reality, and so is actual *simpliciter*; (2) The future is not actual as of the present moment. And Oaklander asks what the second of these claims means. He proposes two possible interpretations. The first is that events that occur at a later time do not occur at the present time. The second is that "as of the present moment the future is excluded from the sum total of reality." He then points out that the first interpretation cannot be right, since on that interpretation nothing is being said that is not true on a static view of time, while, on the second interpretation, one is saying something that contradicts the thesis that the future is actual *simpliciter*.

Oaklander is right that neither interpretation will do. But when I say that the future is not actual as of the present moment, I am not advancing either the trivial claim that states of affairs that are future are not present, or the incoherent claim, discussed earlier, that what is actual *simpliciter* can be different at different times. I am simply employing the temporally-indexed concept of being actual as of a time – a concept that I, in contrast to Oaklander, and also, as we shall see later, to Quentin Smith – believe is a perfectly intelligible concept.

2.6 Truth at a Time and Truth Simpliciter

Finally, Oaklander questions the coherence of corresponding views that I advance concerning truth *simpliciter* and truth at a time. I argue that, for a variety of reasons, a dynamic view requires not just the concept of truth at a time, but also the concept of truth *simpliciter*. (And in any case, if there is such a thing as the totality of what is actual *simpliciter*, then one can define, in a familiar recursive way, the concept of being true *simpliciter*.) It then follows, of course, from my view that, while the future is not actual as of the present moment, it is actual *simpliciter*, that a sentence such as "Gore will be President in the year 2005" is either true or false *simpliciter*. This leads to Oaklander's advancing the following objection:

> But how can, say, "Gore will be President" be true or false *simpliciter*, and also have an indeterminate truth value as of October 7, 1997? If it is true or false *simpliciter*, then it must be so in virtue of some tenseless fact to which it corresponds, but if such a fact exists *simpliciter*, then how can it fail to make true the judgment made on October 7, 1997, that "Gore will be President?" I don't think that there is a consistent set of answers to these vexing questions. (p. 11)

What is happening here is that Oaklander is interpreting the concept of being true at a time as being a matter of correspondence with the tenseless facts that are actual *simpliciter at a given time*. And my response is that, once again, this is an incoherent concept: one cannot temporally index the concept of being actual *simpliciter*. What one has to do is to introduce an *independent* temporally-indexed concept of being actual as of a time, and then define truth at a time in terms of correspondence with what is actual as of a time. If, as Oaklander thinks, such an independent notion of being actual as of a time is not available, then one cannot make sense of the idea of truth at a time in this way. But if Oaklander is right about this, then we have an objection to *any* dynamic view of time, rather than an objection that tells only against the particular dynamic view that I am defending. And, as will be clear from my earlier comments, I see no reason to agree with Oaklander's conviction that the idea of being actual as of a time is unintelligible.

3. RESPONSE TO QUENTIN SMITH'S COMMENTS

3.1 Dynamic Worlds and the Concept of Actuality as of a Time

Most of the first half of Quentin Smith's paper is concerned with the temporally-indexed notion of actuality, and the thrust of his discussion appears to be that the only way that one can make sense of that concept is by analyzing it in *tenseless terms*:

> I suggest that what Tooley claims to be a primitive notion of actuality as of a time is really the familiar notion in the tenseless theory of time that is expressed by "being simultaneous with or earlier than a time," taken in conjunction with a decision to mentally or conceptually abstract from what is later than that time. (p. 25)

My first comment here is that I was somewhat surprised to find a tenser advancing precisely the sort of objection that J. J. C. Smart sets out, for example, in his paper "The Reality of the Future." But, you may ask, why should I be taken aback when a tenser agrees with the tenseless people? After all, I do it myself until the cows come home – indeed, to such an extent that many people may well suspect that I am really a tenseless wolf in tensed sheep's clothing.

The problem, however, is that the concept that is in question here is essential if one is to make sense of the idea of a dynamic world. A dynamic world is precisely one where the facts that there are as of one time may very well differ - indeed, must differ - from the facts that there are as of any other time. To hold that one cannot make sense of the concept of being actual as of time *t* unless one analyzes it as follows

"*E* is actual as of time *t*"

means the same as

"Either *E* is simultaneous with *t* or *E* is earlier than *t*"

really implies, therefore, that the whole idea of a dynamic world is incoherent.

Secondly, Smith himself frequently talks about facts changing from one time to another. Thus, in later criticizing an argument of mine for the view that presentness and pastness cannot be intrinsic properties of events, he says:

> Why can't the defender hold that there is a third way, that with the passage of time some intrinsic temporal properties change, specifically, that when time passes an object can change from possessing the intrinsic property of presentness to possessing the intrinsic property of pastness? (p.27)

But if an object can change from possessing the intrinsic property of presentness to possessing the intrinsic property of pastness, then what facts there are is time-dependent. I do not really see, then, how Smith can fail to accept the idea of being actual as of a time.

Thirdly, what I am doing in Chapter 2 is not, contrary to what Smith appears to suggest, trying to show that the concept of being actual as of a time *must* be primitive. Here, for example, is what I say near the beginning of section 2.2 of *Time, Tense, and Causation*

> A number of defenders of tenseless accounts of the nature of time have suggested, in effect, that no satisfactory account can be given of the concept of being actual as of a given time, that the concept is probably unintelligible, and that, therefore, the whole idea of a dynamic account of the nature of time must be rejected.[1] The best response to this objection would be to provide an analysis of the concept of being actual as of a given time. It seems to me very doubtful, however, that that can be done. What I shall do in this section, accordingly, is to argue that there is nothing objectionable about the view that the concept of being actual as of a given time must be taken as primitive, and as incapable of being analysed. (1997, p. 37)

What I am doing, accordingly, is engaging in a defensive operation. For it is clear that many advocates of tenseless views of time - such as J. J. C. Smart, and Nathan Oaklander - think that the idea of being actual as of a time is incoherent unless it is given a tenseless analysis of the sort suggested by Smith. My strategy, accordingly, was to argue that no illuminating and sound analysis can be given of the idea of being actual *simpliciter*, and thus that it is not a serious objection to a dynamic view of the world if, in setting out such a view, it also turns out that one cannot give an illuminating analysis of the concept of being actual *as of a time*. (One can say in the case of actuality *simpliciter* that A is actual if there exists (tenselessly) an x that is identical with A, but I hold that that analysis is not really illuminating, since one has to understand what existential quantification is, and what the quantifier in question ranges over. But what account can one offer with respect to the latter issue – unless, like David Lewis, one is a "modal realist" – other than that the quantifier ranges over everything that is actual?)

My goal in Chapter 2 of *Time, Tense, and Causation*, in short, is not – as Smith seems to suggest – to prove that the concept of being actual as of a time *must* be analytically primitive; it is rather to argue that if one does take it as primitive, that fact does not constitute a serious objection to a dynamic view of time. Consequently, if it turned out that an illuminating analysis could be given of the concept of being actual as of a time, and one which did not reduce it to tenseless notions, I would be very pleased indeed: that would be the ideal outcome.

Fourthly, one needs to ask what Smith's own view is of the concept of being actual as of a time, since he seems to be suggesting that either it reduces to purely tenseless concepts, or else it is unintelligible. But perhaps his thought is that it can be analysed in terms of *tensed* concepts? Can one, for example, give the following analysis:

> "*E* is actual as of time *t*"

means the same as

> "Either *E* lies in the present at time *t*, or *E* lies in the past at time *t*?"

The objection, however, not only to this analysis, but to any tensed analysis, is that it is *not neutral* between different *dynamic* views of the world. Dynamic views share the idea that facts are time-dependent, and the point of the concept of being actual as of a time is to provide one with a way of talking about that time-dependence. But if one adopted, for example, the tensed analysis just mentioned, then it would follow that the events that are actual as of a given time are *by definition* events that lie either in the present or in the past. Many dynamic views would therefore be ruled out by the supposed analysis of the concept of being actual as of a time – including (a)

Storrs McCall's branching-futures view, (b) classical presentism, and (c) the view that past, present, and future events are all real, but that there is a moving Now.

To sum up:

(1) The concept of being actual as of a time is crucial for dynamic views of time.

(2) If the concept of being actual as of a time cannot be primitive, and can only be analyzed in tenseless terms, then the very idea of a dynamic world is incoherent.

(3) The concept of being actual as of a time must be neutral between different tensed views. It is therefore very hard to see how it can be analyzed in terms of tensed concepts.

3.2 Truth Conditions for Complex Tensed Sentences

Quentin Smith points out that I agree with his claim that the account that Mellor gives of the truth conditions of complex tensed sentences – that is, sentences where one has one tensed operator nested inside of another tensed operator – cannot be correct. (The problem with Mellor's account of such sentences is that his account is in terms of either the existence, or at least the "counterfactual existence," of a token of some sentence at the time that is, so to speak, ultimately being referred to. So if one considers a sentence such as –

"Once upon a time there were no sentences"

– on Mellor's account that sentence is true provided that, had there been a token of the sentence "There are no sentences" back at the time in question, it would have been true. But such a token would, of course, render itself false.)

As Smith points out, I go on to argue that the problem with Mellor's account is easily corrected by employing tensed sentences that do not contain indexicals – sentences such as "Event *E* lies in the present at time *t*" – and so I conclude that Smith's objection cannot be sustained. Smith is not convinced by this response. He does not, however, attempt to show that the analysis I propose for complex tensed sentences generates, in any particular case, the wrong truth value. What he does, instead, is to criticize me on another issue – namely, the view, which I share with advocates of a static view of the world, that ordinary tensed sentences involve indexicals.

3.3 Ordinary Tensed Sentences as Containing Indexicals

If a sentence contains an indexical term, then what proposition it expresses will depend upon the context in which it is uttered. Smith holds that a sentence such as "It is raining in London" always expresses the same proposition, and so he must hold that it does not contain any indexical term.

How can different utterances of the sentence "It is raining in London" express the same proposition? If I utter the sentence on Tuesday, am I not saying something different from what I say if I utter it on Wednesday? For if I were saying the same thing, then someone who uttered the sentence "It is *not* the case that it is raining in London" on Wednesday would be contradicting what I said on Tuesday, and surely that is not so.

What account does Smith give? What he says is this:

> It is consistent to hold that this sentence always expresses the same proposition, namely, that *some event of there being rain in London has the property of presentness*, and yet that the rules of use of this sentence imply that the time that is ascribed presentness by this proposition is whatever time is the time at which the proposition is expressed by the sentence-utterance. (p. 28)

But if "the time that is ascribed presentness by this proposition is whatever time is the time at which the proposition is expressed by the sentence-utterance," then in uttering the sentence "It is raining in London" at different times, one is ascribing presentness to *different times*, so how can it possibly be the case that one is saying the same thing, given that one is referring to one time in one utterance, and to a different time in the other utterance? If a difference in the referents does not make for a difference in a proposition, it is hard to see what does.

3.4 Quentin Smith's Objection to Indexical Analyses

Smith claims that the view that ordinary tensed sentences involve indexicals and that the truth conditions of tokens of such sentences can be formulated in terms of the time at which the tokens were uttered can be refuted by the arguments that he has advanced in Chapter 2 of *Language and Time*. To see whether that is so, consider the truth-conditional analysis of ordinary tensed sentences that I offer:

> Any utterance, or inscription, at time t^*, of the sentence "Event E is now occurring" is true (false, indeterminate) at time t

if and only if

> It is true (false, indeterminate) at time t that E lies in the present at time t^*. (1997, pp. 200-201)

It is important to understand this analysis correctly. First, given that it is an analysis, the statement in question expresses a necessary truth. Secondly, t^* is a variable, and the analysis involves universal quantification over all of the rest of the analysis. So to eliminate possible ambiguity, let me make both of these things explicit:

> It is necessarily the case that, for any time t^*, if U is an utterance or inscription, occurring at time t^*, of the sentence "E is now occurring," then U is true (false, indeterminate) at time t if and only if it is true (false, indeterminate) at time t that E lies in the present at time t^*.

Now, for simplicity, let us replace this by a variant in which truth at a time is replaced by truth *simpliciter*:

> It is necessarily the case that, for any time t^*, if U is an utterance or inscription, occurring at time t^*, of the sentence "E is now occurring," then U is true if and only if E lies in the present at time t^*.

How does Smith attempt to show that such analyses are incorrect? His basic strategy involves the idea that the very utterance, U, in question, of the sentence "Event E is now occurring" could have occurred at some time other than time t^*, in which case it would not have the truth conditions given. What Smith is claiming, accordingly, is that the following counterfactual sentence follows from the analysis offered:

> If utterance U had occurred at some time t^{**} distinct from time t^*, then U would have been true if and only if E lies in the present at time t^*.

But this does not follow from the analysis that I offer. For suppose that U, which in the actual world, occurs at time t^*, had occurred at a different time t^{**}. What would be true? Since the analysis expresses, as all analyses do, a necessary truth, it must obtain in the counterfactual world. But the analysis, when combined with the counterfactual assumption that U occurs at time

t^{**} entails that *U* would then be true if and only if *E* lies in the present at time t^{**}. It does *not* entail that *U* would then be true if and only if *E* lies in the present at time t^{*}. So the upshot is that my analysis generates the right truth values for counterfactual worlds where a given utterance occurs at a different time than in the actual world.

Let us now consider, however, how Quentin Smith's overall discussion in Chapters 2 and 3 of his book *Language and Time* bears upon the account that I have offered of tensed sentences. In Chapter 2, Smith advances a plethora of arguments in support of the thesis that ordinary tensed sentences cannot be translated by means of what he calls tenseless date-sentences, Then, in Chapter 3, he argues that ordinary tensed sentences cannot be translated by means of tenseless token-reflexive sentences. And in his comments earlier, Smith pointed out that I do not, in my book, respond to any of those arguments, and he went on to ask: "Given this, with what justification could Tooley deny the claim that a refutation of his theory appears in Chapter II of my *Language and Time*" (p. 7)?

But how do those arguments bear upon my account? For those arguments are concerned with the *translation* of ordinary tensed sentences into *tenseless* sentences, whereas, first of all, I hold that *no translational* analysis can be given for ordinary tensed sentences, and secondly, the sentences that I use to specify the truth conditions of ordinary tensed sentences are themselves *tensed* sentences, of another type - namely, tensed sentences that involve no indexicals.

Smith's line of thought, I suspect, is that any account of the truth conditions of ordinary tensed sentences that, as he would put it, uses dates, is open to a decisive objection. But if so, he is mistaken, for the reason indicated above. But let me point to the exact place in *Language and Time* where the error occurs. On pages 36-37, Smith discusses a view advanced by J. J. C. Smart, and he refers there to the following sentence:

> (4) "*E* is present" as spoken by *P* at *t* in world *W* is true in *W* if and only if *E* is at *t* in *W*. (Smith (1993), p. 37)

Smith's comment is then as follows:

> I agree with Smart that (4) is true, but this fails to save the date-sentences theory, since the fact that the non-world-indexed version of (4) is false [that is," "*E* is present" as spoken by *P* at *t* is true if and only if *E* is at *t*"] suffices to refute the theory that date-sentences give the meaning of A-sentence tokens. World-indexed truth conditions do not give the semantic content of utterances, since the criterion of truth for those sentences is that the clause on

> the right side of the biconditional that has the world-indexed phrase appended to it has the same truth value in the mentioned world as the utterance denoted by the clause on the left side of the biconditional. This allows any clause with the same truth value in *W* as 'E is at *t*' to be substituted for 'E is at *t*' and this prevents (4) from giving the meaning of the utterance. (1993, p. 37)

This passage involves the two errors that I pointed out above. First, the statement that gives the truth conditions for an utterance of "*E* is present" must be viewed as involving universal quantification. So the non-world-indexed version of (4), formulated explicitly, is

> (4*) *For any person P, and any time t,* "*E* is present" as spoken by *P* at *t* is true if and only if *E* is at *t*.

And as we saw above, such a universally quantified statement does not entail that the relevant counterfactuals about what would be the case if *E* had occurred at a different time get assigned the wrong truth values.

Secondly, any statement that expresses an analysis must be a necessary truth, so it is not true that one can, as Smith suggests, replace any clause in the analysis by any clause that merely has the same truth value.

Finally, however, there is another reason why I did not discuss the arguments in the first part of *Language and Time*, and that is that the basic approach that Smith follows there is, I believe, unsound. For consider his approach. In Chapters 2 and 3, he argues that various proposed translations of ordinary tensed sentences into tenseless sentences are unsound, and the general conclusion that he tries to establish is this:

(1) For any token *S*, of an ordinary tensed sentence, there is no tenseless sentence *T* (either a sentence-type or a sentence-token) which is such that *S* and *T* have the same truth condition.

In addition, he believes - as he in effect indicates on page 12 - that this conclusion can be extended so that one can also refute accounts of the truth conditions of ordinary tensed sentences that, while *specifying* the truth conditions of tensed sentences, do *not* provide *translations* for them. In short, Smith thinks that he can move from (1) to

(2) For any token *S*, of an ordinary tensed sentence, there is no tenseless sentence *T* (either a sentence-type or a sentence-token) which is such that *T* gives the truth conditions of *S*.

How does Smith think that one can get from (1) to (2)? One idea would be that (2) follows from (1) because:

(3) If *T gives* the truth conditions of *S*, then *T has* the same truth conditions as *S*.

But (3) is clearly false. For consider:

(*A*) "Schnee ist weiss."
(*B*) " 'Schnee ist weiss' is true if and only if snow is white."

B gives the truth conditions of *A*, but *B* and *A* do not have the same truth conditions, since the second sentence, but not the first, can be true even if snow is *not* white.

Sentence *B* does, however, contain a part – "snow is white" – that is synonymous with "Schnee ist weiss" – and so one might try getting from (1) to (2) via:

(4) If *T gives* the truth conditions of *S*, then *either T, or some part of T*, has the same truth conditions as *S*.

But (4) is also false. For consider some token of "I am in Chicago" – call it *C* – that I have uttered. The truth conditions of that token are given by:

(*D*) *C* is true if and only if there is some *x* such that the occurrence of the pronoun "I" which is part of *C* refers to *x*, and *x* is in Chicago.

But though *D* gives the truth conditions of *C*, *D* does not have the same truth conditions as *C*, nor does *D* contain a part that has the same truth conditions as *C*. So (4) is false.

The conclusion here is general, and it is as follows: there is no way of getting from (1) to (2), and the reason is simply that although no non-indexical sentence *has* the same truth conditions as any token of any indexical sentence, non-indexical sentences can be used to *specify correctly* the truth conditions of tokens of indexical sentences. The upshot, accordingly, is that Smith's arguments do not go through.

Another way of seeing this is by noticing that, if those arguments were sound, they would prove too much, since Smith's arguments against the possibility of stating truth conditions for ordinary tensed sentences in tenseless terms can be exactly paralleled by arguments against the possibility of stating truth conditions for ordinary *spatially "tensed"* sentences that is, sentences containing terms such as "here" and "there" – in spatially "tenseless" terms – that is, by means of sentences that refer to spatial locations, but that do not involve any reference to the "spatially present" concept of hereness.

The moral here is a very general one. Whenever one advances an argument for the conclusion that one cannot give tenseless truth conditions for tensed sentences, one needs to check to see if the argument can be paralleled in the spatial case. If it can be, then something is wrong with the original argument.

What anyone who wants to show that one cannot give tenseless truth conditions for tensed sentences really needs to do, accordingly, is to set out an analysis of sentences containing terms like "here" and "there," and then to show that the sort of analysis that works in that case does not work in the case of tensed sentences. Nowhere in Quentin Smith's *Language and Time* does one find an argument of that sort.

3.5 Quentin Smith's Move from Analysis to Metaphysics

In *Time, Tense, and Causation*, I said in section 8.8.2 that metaphysical argument is needed in the philosophy of time, and that, in particular, one could not refute a static view of time simply by showing that no adequate tenseless account can be offered for the truth conditions of tensed sentences. Smith, in his comments, suggests that I failed to address an argument that he advanced in section 1.4 of *Language and Time*. What is true is that I did not discuss his argument in a detailed way; however it was in fact Smith's argument that I especially had in mind when I said:

> Or suppose that there was some decisive objection to any tenseless analysis of tensed sentences. It would then follow that everyone who used tensed sentences, and interpreted them in the standard way, was expressing the belief that the world is dynamic. But this does not take one very far, since, while the fact that a belief is widely shared provides one with some reason, other things being equal, for thinking that there may well be good reasons to accept that belief, one still has to ask whether the belief is really justified, and if so, how. (1997, p. 248)

Let's now take a closer look, however, at the relevant argument. At the heart of it is the idea that the following inference is sound:

> (6) Ordinary *A*-sentences ascribe *A*-properties, some tokens of these sentences are ordinarily accepted as true, and there are no logical or scientific reasons to reject all these sentences as false. (Smith (1993), p. 16)

Therefore:

> (7) Some tokens of ordinary *A*-sentences are true, and events possess *A*-properties *(ibid)*.

What is one to say about this pattern of inference? Consider some parallel arguments. First, concerning libertarian free will:

(6a) Ordinary libertarian-free-will-sentences ascribe libertarian-free-will-properties to actions, some tokens of these sentences are ordinarily accepted as true, and there are no logical or scientific reasons to reject all these sentences as false.

Therefore:

(7a) Some tokens of ordinary libertarian-free-will-sentences are true, and actions possess libertarian-free-will-properties.

Secondly, concerning objective values:

(6b) Ordinary objective-value-sentences ascribe objective-value-properties to states of affairs, some tokens of these sentences are ordinarily accepted as true, and there are no logical or scientific reasons to reject all these sentences as false.

Therefore:

(7b) Some tokens of ordinary objective-value-sentences are true, and states of affairs possess objective-value-properties.

Thirdly, concerning souls, or demons, in a slightly more medieval society:

(6c) Ordinary soul-sentences ascribe soul-properties to humans, some tokens of these sentences were ordinarily accepted as true, and there were, at the time in question, no logical or scientific reasons to reject all these sentences as false.

Therefore:

(7c) Some tokens of ordinary soul-sentences were true, and humans possessed soul-properties.

(6d) Ordinary demon-sentences ascribe demon-properties to entities, some tokens of these sentences were ordinarily accepted as true, and there were, at the time in question, no logical or scientific reasons to reject all these sentences as false.

Therefore:

(7d) Some tokens of ordinary demon-sentences were true, and some entities possessed demon-properties.

What is one to say about such arguments? It seems to me that what one should say is that all of them are bad arguments. For what Smith's argument and these parallel arguments all involve is what might be called an "innocent until proven guilty" approach to epistemology: if some beliefs are rather widely accepted, and there is no evidence against them, then the beliefs are justified. This principle seems to me clearly unsound. For one thing, there may be any number of *logically incompatible* beliefs that there is no evidence against, and more than one of those beliefs might be quite widely accepted: imagine, for example, different and incompatible beliefs about how many deities there are who never intervene in the physical world. Do we have logical or scientific evidence against the claim that there are exactly three such deities? Or exactly two? If not, then the principle that underlies Smith's argument would entail that all such beliefs that are sufficiently widely accepted are justified. I conclude, therefore, that the principle in question should be rejected.

Smith seems to view the person who rejects his argument here as defending some sort of skeptical position. But the arguments we've just considered make it clear that a rejection of this type of reasoning has nothing to do with skepticism.

How should one approach these issues? As a foundationalist, I would claim. Some beliefs are non-inferentially justified, and other beliefs are justified on the basis of evidence. In neither case does the mere absence of negative evidence provide grounds for thinking that beliefs that are ordinarily accepted are likely to be true. *Positive* justifying grounds are needed.

Might one attempt to revise Smith's argument by noting that demons and souls are not the sort of things with which one can be *directly acquainted*, and then by arguing that the situation is different in the case of tensed properties, and that, because of that difference, Smith's argument is acceptable, whereas my formally parallel arguments are not?

Such a revision seems to me unpromising, for two reasons. In the first place, there are good reasons for saying that one is not directly acquainted with most tensed properties. In the case of the property of futurity, if events do not exist until they are present, then there is no property of futurity that events sometimes have. But even leaving aside that controversial piece of metaphysics, there is the fact that we do not have non-inferentially justified beliefs about future events - something that we would have to have if we were directly acquainted with the property of futurity.

The same seems to me to be true of the property of pastness, though the argument is slightly more subtle here, since some epistemologists would claim that we have non-inferentially justified beliefs about the past. But even if one does not challenge that contention, there is another argument, which turns upon the idea that one cannot be directly acquainted with a property of some state of affairs unless one is directly acquainted with the state of affairs itself, and all of its constituents. The point is then that when I remember seeing a red apple, I am not directly acquainted with redness. So even if my belief that I saw a red apple is non-inferentially justified, it cannot involve direct acquaintance with pastness since it does not involve direct acquaintance with the redness which must be a constituent of the state of affairs to which the pastness attaches.

There is, however, a second reason why an appeal to direct acquaintance is not a way to attempt to resurrect Quentin Smith's argument - namely, that if one could show that it was plausible that one was directly acquainted with tensed properties, no further argument would be necessary.

3.6 Absolute Simultaneity

Finally, Quentin Smith and I share an enthusiasm for the idea that events in the world can stand in the relation of absolute simultaneity, and Smith views with more favor than Storrs McCall my invocation of quantum mechanics. Smith asks, however, whether I shouldn't have "ascribed to the standard neo-Lorentzian thesis that motion through absolute space (or some medium in absolute space) is the cause of time dilation and length contraction" (p. 29). The answer is that, in fact, I find that view very attractive, and my reason for not mentioning it was simply that I wanted to focus my discussion instead upon the setting out of a metaphysical argument for absolute simultaneity - that is, the argument mentioned earlier, and which is based on a consideration of alternative hypotheses concerning causal relations between spacetime points.

NOTES

1 A forceful formulation of this view can be found in J. J. C. Smart (1981).

REFERENCES

Broad, C. D. (1923), *Scientific Thought* (London: Routledge and Kegan Paul).
McCall, S. (1994), *A Model of the Universe* (Oxford: Oxford University Press).
Smart, J. J. C. (1981), "The Reality of the Future," *Philosophia* 10: 141-150.
Smith, Q. (1993), *Language and Time* (New York: Oxford University Press).
Stein, H. (1968), "On Einstein-Minkowski Space-Time," *Journal of Philosophy* 65: 5-23.
Tooley, M. (1997), *Time, Tense, and Causation* (Oxford: Oxford University Press).

5
MCTAGGART, CHANGE AND REAL TENSE: A CRITICAL NOTICE OF HUGH MELLOR'S *REAL TIME II*

Mark Hinchliff
Reed College

McTaggart put his finger on the respective weaknesses of the A and B theories of time: the B theory apparently cannot account for the possibility of change; the A theory is apparently inconsistent. Since McTaggart, A and B theorists have attempted to shore up the weaknesses of their respective views and to attack the weaknesses of their opponent's views.

Hugh Mellor's *Real Time* has been since its publication in 1981 the canonical defense of the B theory. His *Real Time II* (1998) strengthens that defense and integrates it with his subsequent work in metaphysics.

Mellor's attack on the A theory remains the same from *Real Time* to *Real Time II* – an update of McTaggart's attack. However, Mellor's views on change have undergone a significant revision.

My talk has two parts. In the first I argue that Hugh's attack on the A theory fails because it relies on two principles that are not obviously true, are definitely question begging, and quite likely false. In the second part, I argue that Hugh's new account of change still leads him to deny an intuitive and essential feature of change.

I. MCTAGGART

I stand accused by Hugh of a "capital offence" (Mellor (1998), p. 72). According to Hugh, I am guilty beyond a reasonable doubt of self-contradiction. The proof of my guilt comes from McTaggart. "However," Hugh writes,

> the debate is too well worn by now to be settled simply by repeating McTaggart's proof, sound though it is. To change the metaphor, too many people have managed to inoculate themselves against it. So if it is to wipe out belief in A–facts, as it should, we need a more virulent strain of it, a strain which I think is best nurtured on the B-facts that make A–propositions true or false. (1998, p. 75)

L. N. Oaklander (ed.), The Importance of Time, 59–67.

So here I sit, watching time's river flow, and Hugh is trying to wipe me out with germ warfare. I had better respond.

Hugh makes use of truthmakers in his argument and throughout his book. I'm yet to be convinced of the value of truthmakers in metaphysics, but I understand them well enough to follow the argument. Truthmakers are supposed to provide the answers to questions of the form, What makes S true?, by connecting truth with being. For example, what would make the sentence "e is earlier than f" true would be the *existence* of the truthmaker that e is earlier than f, and what would make it false would be the non–existence of that fact.

An A-theorist who believes in truthmakers, which for the sake of discussion I will henceforth take myself to be, will then hold that all tokens of a sentence like "e is past" are made true by a single fact like the fact that e is past. When that fact exists, all tokens of "e is past" are true; when that fact doesn't exist, those tokens are all false.

It is this principle against which Hugh launches his attack. He describes a case in which this principle is false, a case in which two tokens of "e is past" have different truth values at the same time. These tokens cannot both be made true by a single fact. Since A-facts like the fact that e is past are introduced solely for the purpose of making tokens of A-sentences like "e is past" true and they do not serve that purpose, we have no reason, Hugh concludes, to believe in the existence of these facts or, by implication, in the A-theory itself.

It all comes down to Hugh's case. Let us suppose that *e* is Jim's race, and Jim's race is at 4:30. Let *a* be a token of "*e* is past" that occurs at 3:30, and *b* be a token of "*e* is past" that occurs at 5:30. On the A-theory, a and b are made true by the A-fact that e is past. So at 4 o'clock, when the fact that e is past does not exist, a and b are both false. And at 5 o'clock, when the fact that e is past does exist, a and b are both true.

According to Hugh, these truth values are "obviously wrong:"

> To say *before* Jim's race that it is past is to produce a token that is and always will be false. Similarly, to say *after* his race that it is past is to produce a token that is and always was true. ... These must be the right truth values [Hugh goes on]. Once we distinguish propositions from their tokens, it is obvious that tokens of an A-proposition, unlike the proposition itself, do *not* change their truth value over time. (1998, p. 78)

I do not find these truth values wrong. I certainly do not find them to be obviously wrong. In fact, I find the matter to be quite subtle. Several issues seem to be in play.[1] I'll focus on the main one here.

Hugh appeals to two principles in his argument:

(1) If a token is true (false), it always will be true (false);
(2) If a token is true (false), it always was true (false).

It is the conjunction of these principles which allows Hugh to conclude that a token of "e is past" does not change its truth value over time. However, these principles are not the truisms Hugh takes them to be, though they are very close to such truisms. To see their unobviousness better, let us replace in the principles talk about the truth values of tokens with talk about earthquakes. We thus obtain

(1E) If there is an earthquake, there always will be an earthquake;
(2E) If there is an earthquake, there always was an earthquake.

These principles are not in general true. They jointly entail that if there is ever an earthquake, there is always an earthquake. But if the earth is quaking, it doesn't have to keep quaking and it doesn't have to have been always quaking. Truisms about earthquakes, however, lie close to hand:

(1E*) If there is an earthquake, it will always be that there was an earthquake;
(2E*) If there is an earthquake, it has always been that there would be an earthquake.[2]

Moreover, these truisms about earthquakes correspond to truisms about the truth values of tokens:

(1E*) If a token is true, it will always be that it was true;
(2E*) If a token is true, it has always been that it would be true.

These obvious truths lend no support to Hugh's conclusion that a token does not change its truth value over time. And without that conclusion, Hugh cannot construct his case against the A-theory of two tokens of the same sentence differing in truth value at the same time.

The principles which do yield that conclusion, we now see, are not obvious. And once we clearly distinguish them from the principles which are, we have no reason for thinking them true, every reason for thinking them question begging, and some reason, given their similarities to (1E) and (2E), for thinking them false.

We are all familiar with stories, both real and imagined, in which a deadly germ differs ever so slightly to our great horror or relief from an innocent organism. I am relieved to discover that Hugh has released upon us not a deadly plague but a harmless gas.

II. CHANGE

It is common ground that a candle changes its shape if it has incompatible properties at different times. It is also common ground that the sentence "the candle is straight at *t*" expresses a relation between the candle and *t*. The presentist and the friend of temporal parts will analyze this relation in terms of the property of being straight. The friend of temporal parts will analyze it as follows: the part of the candle that is at *t* has the property of being straight.

Hugh, however, is not a friend of temporal parts, at least not for things that can undergo change. The friend of temporal parts denies our intuition that the candle undergoes a change in *its* shape. As Hugh beautifully put it, "different things differing in their properties do not amount to change even when, as here, one is later than the other, and both are parts of something else" (1998, p. 111). The things that undergo change must be wholly present at each moment of their existence; they must *endure*. But once the candle endures, its shapes, as Hugh clearly saw in *RT*, couldn't be properties for the detenser. The shape of a temporal part can be a property because different parts have the incompatible shapes. But once it is one and the same thing with the incompatible shapes, they cannot be treated as properties without engendering contradiction. The shapes must be relations, relations that one and the same candle stands in to distinct times. The shapes are still incompatible; they are incompatible relations, which no one thing can bear to the same time. Hugh and the friend of temporal parts thus reach a standoff. The friend of parts is able to treat the shapes as properties but for the price of denying our intuition that the candle changes *its* properties. Hugh is able to accommodate this intuition, but for the price of denying our intuition that shapes are properties.

Though Hugh tries to mitigate the unintuitiveness of his position, it is understandable that in *Real Time II* he should offer a new theory of change. Hugh now favors an "operator" analysis of sentences like "the candle is straight at *t*" and "the candle is bent at *t*." On an operator analysis of "the candle is bent at t," the temporal modifier "at t" is an operator and "the candle is bent" is an operand. The temporal modifier is thus like the negation or modal operators. The extensions of operators like these are usually represented by functions from the appropriate semantic values to truth values.[3] In the case of negation, the appropriate semantic value is the truth value of the operand. In the case of the modal operators, it is an intension of some kind. In the case of the temporal modifiers, it is a temporally sensitive content of some sort.

An operator analysis of the temporal modifiers has traditionally been the preserve of A-theorists like the presentists, with Arthur Prior being the pioneer. It is worth saying why. For a presentist like myself, "the candle is bent" is a non–indexical sentence. It is a sentence because it can have a truth value; it can be simply true or simply false. It is true if the candle has the property of being bent, false otherwise. It is a nonindexical sentence because it expresses the same proposition in every context of utterance, the proposition that the candle is bent, a proposition, we may assume, which has the property of being bent as a constituent.[4]

It is this proposition, the proposition that the candle is bent, on which the temporal modifiers operate. It is true at some times, false at others; but – and this is the essential point – it can be simply true or simply false. How can the candle be simply bent when it is bent at some times, not at others? One time would have to be special, distinguished from all the others. And the way things were at that time would have to be the way things simply were. For the presentist, there is such a time – the present: the way things are at present is the way things are. If the candle is presently bent, it is bent. Since the candle could presently be bent, it could simply be bent, and the proposition that the candle is bent could be simply true or simply false.

The presentist doesn't identify the way things are with the way things presently are *as a matter of definition*. Presentism is like actualism in the metaphysics of modality: the way things are is the way things actually are, but not as a matter of defining them to be so. When the presentist says that the candle is bent if and only if it is presently bent, the left side is to be understood to contain no hidden or implicit temporal modifiers. It is to be read as being like the sentences "7 is prime" and "The temporal part of the candle that is at t is bent." The friend of temporal parts and the presentist thus both recognize the property of being bent as a property that a thing can simply have. The friend of temporal parts can do so because the temporal parts couldn't exist at more times than one and thus couldn't be both straight and bent. The presentist can recognize the property of being bent because the way the candle is at present is the way the candle is, and the ways the candle is at other times are not.

We can now see why the operator analysis has traditionally been the preserve of the presentist. Because the B theorist does not think one time is distinguished from all the others, the B theorist cannot infer from the fact that the candle is bent at one time that the candle is simply bent, without also inferring from the fact that the candle is straight at another time that the candle is also straight, which leads to the contradiction that the candle is both bent and straight (a capital offense in some parts).

The B theorist must therefore deny that we can ever infer that the candle is bent from the fact that it is bent at a time. We can therefore never say that the candle is bent, only that the candle is bent at some time or other. Being bent is thus not a property; it is not something that the candle can simply have. The proposition that the candle is bent, with the property of being bent as a constituent, is thus not available for the B theorist. So it cannot be what the temporal modifiers operate on.

Hugh is, of course, aware of these reasons, or of these sorts of reasons, for thinking that the B theorist cannot give an operator analysis of sentences like "the candle is bent at t" which captures our intuition that the shapes are properties. It is through these treacherous waters that Hugh attempts to sail in *RTII.*

What I shall argue is that though Hugh succeeds in giving an operator analysis of sentences like "the candle is bent at t," he still fails to capture our intuition that the shapes are properties. Giving an operator analysis is not to be equated with capturing our intuition. One can give the one without capturing the other.

Let us introduce the notion of a *relativized property*. A thing can have a relativized property only relative to something else, for example, a world, a time, or a place. A relativized property isn't a property because a thing couldn't just have it; it could only have it relative to something else. A relativized property also needs to be distinguished from a relational property, such as being six miles from the equator. A relational property is a property - a thing can simply have it. It is relational because it is composed out of a relation, such as being *six miles from.*

If we treat being bent as a relativized property, we can treat the candle's being bent as a *relativized proposition*, which can have a truth value only relative to something else, in this case, a time. A relativized proposition is no more a proposition than a relativized property is a property. It cannot be true or false, only true or false relative to something else. The relativized proposition that the candle is bent is true at a time only if the candle has the relativized property of being bent at the time. The temporal modifiers can then be taken to operate on these relativized propositions. In this way, one can have an operator analysis of sentences like "the candle is bent at t" without capturing our intuition that the shapes are properties.

Though Hugh does not put his analysis in these terms, it is, I believe, the analysis he gives. In *RTII*, Hugh sees two obstacles he must overcome if he is to offer an operator analysis that treats shapes as properties. The first obstacle we have already encountered: the expression "the candle is bent" fails to express a proposition on the B theory, but the temporal modifiers are supposed to be prefixed to sentences that express propositions (1998, p. 91).

Hugh replies as follows:

> But this need not be so. Consider for example the view that there are many possible futures, none of which is yet actual, and the operator "It is possible that" prefixed to a future sentence "P" like "Jim races tomorrow." On this view "It is possible that P" will be true even though "P" lacks a truth value. Similarly, if for different reasons, "at t, Fa" may be true even if "Fa" lacks a truth value. (1998, p. 94)

This reply is fine as far as it goes. A B theorist like Hugh can offer an operator analysis of reports of change, as we have seen. But as we have also seen, offering an operator analysis isn't by itself sufficient to capture the intuition that shapes are properties. The second obstacle and Hugh's response move us closer to determining whether Hugh finds a way to admit the shapes as properties.

Here is how he puts the second obstacle. The expression "the candle is bent" does not have a single truth value, because it has different truthmakers at different times. There doesn't seem to be the truthmaker that the candle is bent located at various times. There seem to be many truthmakers, such as that the candle is bent at t, which are not located but have the times as constituents. But with the times as constituents of the truthmakers, the shapes seem to be relations holding between the candle and the various times. (This paragraph is a paraphrase drawn from (1998, p. 94).

I would choose to put this problem slightly differently. Hugh suggests that "the candle is bent" has "no one truthmaker" and "no one truth value" because it has *many* different truthmakers. But of these many different truthmakers - that the candle is bent at t, that the candle is bent at t', and so forth - some could exist, and some could not. If some exist and some do not and they are truthmakers for "the candle is bent," then it is both true and false. So it seems better to say that "the candle is bent" has no one truthmaker and no one truth value because it has no truthmaker and no truth value. The problem for Hugh is to find a truthmaker for "the candle is bent."

In reply, Hugh distinguishes a truthmaker's having a time as a *constituent* from its having a time as a *location*. The relational truthmaker that the candle is bent at t has the time as a constituent, whereas the nonrelational truthmaker that the candle is bent, which is located at t, does not have the time as a constituent but as a location.

On Hugh's view then, at every time at which the candle is bent, there is located a truthmaker that the candle is bent. And at every time at which the candle is straight, there is located a truthmaker that the candle is straight. Because these truthmakers are only located at the various times and do not have the times as constituents, the shapes which they do have as constituents are not relations to the times but properties of the candle. A B theorist, Hugh will say, can thus give an operator analysis of the reports of change which captures our intuition that the shapes are properties.

I naturally have several questions to ask. Perhaps the most pressing concerns whether the shapes are properties on Hugh's theory. Let us take a truthmaker that the candle is bent which is located at t and ask this question: Does it exist? By this question I do not mean to ask whether it exists at t. I assume the answer to that question is, Yes. My question is analogous to a question we might ask about Bill Clinton: Granted that he isn't here, at this place, does he exist? The answer is, Yes.

If the answer to my question about the truthmaker is also, Yes, that is, if the truthmaker exists, then the expression "the candle is bent" is true. However we must give the same affirmative answer to the question, Does the truthmaker that the candle is straight, which is located at t', exist? But if it exists too, then the expression "the candle is straight" is true. But the candle cannot be both straight and bent; those shapes are incompatible.

These two truthmakers cannot both exist. And if one exists, so must the other. The B theory gives us that: differences in temporal location are like differences in spatial location; they do not entail existential differences. It follows that neither of these truthmakers exists.

These truthmakers now seem quite peculiar. They do not exist, but they are located at t and t'. How can something that doesn't exist be located at all? The answer I anticipate is that though these truthmakers do not exist, they do exist at their temporal locations. The truthmaker that the candle is bent doesn't exist, but it exists at t. And the truthmaker that the candle is straight doesn't exist, but it exists at t'.

These truthmakers exist at times and only at times; they do not simply exist. They are *relativized truthmakers*. They correspond to the relativized properties introduced earlier. If the relativized truthmaker that the candle is bent exists at t, the candle has the relativized property of being bent at t. If the truthmaker doesn't exist at t, the candle doesn't have the relativized property. The constituent of the relativized truthmaker is thus not the property of being bent but the relativized property of being bent. Hugh has failed, I think, to capture our intuition that the shapes are properties.

My argument against Hugh's account of change in *RTII* rests on a distinction and an intuition, the distinction between properties and relativized properties and the intuition that shapes are properties. One might question whether the distinction and the intuition can support the weight I put on them. I will close with three remarks on this concern.

The presentist shares the intuition that shapes are properties with the friend of temporal parts, and they understand what a property is in the same way. Second, even Hugh will admit the distinction. The relational property of being bent at t *is* a property; the candle can just have it, unlike being bent, which the candle can have only at a time. Finally, the distinction between properties and *relations* seems clear and well founded, as does the intuition that the shapes are not *relations* but properties. Hugh has that intuition, though he denies it in the end. I make this final point because relativized properties seem little different, if different at all, from relations. I am far from home; I stand in the far-from relation to my home. Or is it that I have the relativized property of "being far," which a thing can have only at a place? In my case, I have the property at my home. Here I think it is clear that the relativized property is simply the relation in disguise. We are not taken in in this case, because we are not predisposed to think that being far is a property. I submit that the situation is no different in the case eof shapes, except that we are predisposd to believe them to be properties. Once we correct for the distorting influence of our belief, however, we see that the relativized properties of being far and being bent are at bottom the same - relations. I thus find the distinction and the intuition to be both clear and robust, and I find a theory of change that denies the intuition to be - well, I wouldn't say, guilty of a capital offense - but to be to that extent unsatisfactory theory.*

*Note added after the session. In the comments I delivered on Hugh's account of change, I thought Hugh was trying to find a truthmaker for the expression "the candle is bent." I objected that this truthmaker would have to be temporally relativized and thus too its constituent property, forcing Hugh to deny our intuition that shapes are properties. I evidently misunderstood Hugh on this point however. As Hugh made clear in his reply, he is looking for a truthmaker not for "the candle is bent" but for "the candle is bent at t," which turns out to be the candle-is-bent fact located at t. Fair enough.

But sometimes you find what you are not looking for. Why isn't the candle-is-bent fact located at t *also* a truthmaker for "the candle is bent?" I understand that Hugh does not *want* "the candle is bent" to have a truthmaker; for if it did, so would "the candle is straight," yielding a contradiction. But what principles of Hugh's theory preclude "the candle is bent" from having as its truthmaker the fact that the candle is bent (a fact which happens to be located at t)? If these truthmakers are like particulars located in space and time (as Hugh suggests), then they would seem to exist in the same sense as Socrates exists for the B theorist. And if a truthmaker for a sentence exists, the sentence is true.

NOTES

[1] Two other matters are tokens of propositions and the truth values of tokens opposed to the truth values of the propositions they express.
[2] I set aside worries about the unreality of the past or future. Such worries are of no help to Hugh.
[3] I omit complications due to indexicals and indexical operators, which require a doubly indexed function.
[4] Assuming that being bent is a property and propositions have constituents. I place no weight on these notions.

REFERENCES

Mellor, D. H. (1998), *Real Time II* (*RTII*) (London: Routledge).

6
COMMENTS ON D. H. MELLOR'S *REAL TIME II*

L. A. Paul
University of Arizona
Australian National University

Real Time II, Professor Mellor's revised version of his classic *Real Time*, presents us with a new and well developed B-theory of time. B-theories of time hold that there are no temporal properties of past, present and future. Instead, there are temporal relations of later than, simultaneous with, and earlier than. Mellor argues that what makes "e is past" true at time t is the fact that event e is located at some time earlier than t, while "e is present" is made true at t by the fact that e is located at t, and "e is future" is made true at t by the fact that e is located at some time later than t.

One of the most interesting aspects of *Real Time II* is the link that Mellor develops between his theory of time and his new theory of causation, argued in detail in his book *The Facts of Causation*. The link enables Mellor to provide strong responses to some of the traditional problems for B-theories. The direction of time is explained by holding that causal order fixes time order (and by arguing that there is a unique causal order), time is distinguished from the dimensions of space because time is "the causal dimension of spacetime," and an account of the nature of change is provided when we see that "the dimension of change must also be the dimension of causation" (1998, p. 6). As we can see, Mellor's views on time depend heavily on his views about causation.

The way in which causation is the workhorse for Mellor's theory of time emphasizes the interest and importance of his theory of causation for philosophers of time. Mellor himself has pointed out that accounts of causation do a lot of work for many extant theories, but that "[a]ll too often crucial but contentious assumptions ... are backed by unargued intuitions. That is not good enough: the more we need to invoke causation, the more we need a full and fully argued account of it to tell us what it is, what it entails, and what it can and cannot do" (1995, p. 3). I agree. Accordingly, I shall focus on Mellor's theory of causation and its connection with some of the problems for B-theories that Mellor hopes to resolve. Since Mellor has

L. N. Oaklander (ed.), The Importance of Time, 69–80.

given us a causal B-theory, problems with his theory of causation may translate into problems for his theory of time.

Mellor's theory of causation is based on chance: causes are facts which raise the chances of their effects. When C and E are facts (i.e., statements or propositions C and E are true), C causes E only if the chance of E if C is greater than the chance of E if ~C. The account makes use of counterfactuals: we are to evaluate the chance of E without C in the closest possible world. For Mellor, if C causes E, then "E because C" is true, and "E because C" is the preferred form for basic causal truths.[1]

There are some problems that I have with Mellor's account of causation, worries which, because of the strong connection Mellor makes between time and cause, carry over to his theory of time. The problems revolve around central, much discussed issues in the causal literature: the handling of redundant causation, the individuation of facts as causes and effects, and the possibility of backwards causation. Because of their important role in the evaluation of any theory of causation, these issues need to be addressed to give Mellor's causation-based theory of time its widest appeal.

The problem involving redundant causation is a general one: it seems to show that Mellor's account of causation, that causes are facts that raise the chances of their effects, gives us neither necessary nor sufficient conditions for causation. If the account gives us an inadequate characterization of causation, then we are left with an incomplete theory of time. The problem of the individuation of facts is directly related to Mellor's account of change. Mellor holds that factual causation is necessary for change and to distinguish temporal variation from spatial variation. But his method of individuating facts, which his account of factual causation relies upon, creates serious problems for his view of causation and thus of change. Finally, Mellor argues that backwards causation is logically impossible. If this claim is correct, it allows him to reject the possibility of circular time and time travel. However, I will argue that Mellor's argument against backwards causation is flawed, or at least incomplete.

1. PROBLEMS WITH MELLOR'S ACCOUNT OF CAUSATION

The first problem for Mellor's account of time involves a problem for his theory of causation involving cases of redundant causation, where we have effects with two or more potential causes. These examples show that Mellor's account of causation, that "E because C" is true if causes raise the chances of their effects, gives us neither necessary nor sufficient conditions for causation. [A note: I do not find Mellor explicitly saying that a fact raising the chance of another fact is sufficient for causation, but if this were not the case, his theory wouldn't count as a "substantive metaphysical theory" of causation (1995, p. 5) nor would chances raising the cause of an effect provide the "real embodiment of causation" (1995, p. 169).]

Two examples will show how the theory does not provide necessary or sufficient conditions for causation.[2] Example one: the late shooter. Two people with guns are pursuing a victim. The first gunman, Fred, reaches the victim, Vicky, and shoots her dead, clean through the heart. Vicky dies the instant the bullet passes through her heart, We may say that Fred's shooting increased the chance of the fact that Vicky dies. Just after Fred shoots, but just before the bullet hits Vicky, Sally (the second gunner) bursts onto the scene and shoots Vicky as well. Sally's bullet passes through the hole in Vicky's heart a split second after Fred's bullet. Sally's shooting also increased the chance of Vicky's death. The case is one of a species of late preemption. Sally's shooting increases the chance of Vicky's death, but is not a cause of it, so increasing the probability of an effect is not sufficient for causation.

For example two, change the case slightly. Make it a case of early preemption, but one in which the preempting cause lowers the chance. This time, Fred is a terrible shot, but Sally won't shoot if she thinks Fred has a chance of killing Vicky. In this case, Sally is with Fred and sees him shoot. Although she has begun pulling the trigger, she refrains from completing the pull. Sally's pulling the trigger and shooting would have increased the chance of the fact of Vicky's death, as before, so Fred's shooting thereby lowers the chance of Vicky's death. Since Fred's lucky shot lowers the chance of Vicky's death but is still the cause of it, increasing the probability of an effect is not necessary for causation.

Menzies (1989) solves cases like the lucky shooter by holding that causation requires continuity, or unbroken causal processes, defining such processes as chains of probabilistic dependences and taking causation to be the ancestral of these processes. Mellor could use a version of this solution as well if he were willing to accept the claim that causation is transitive. [He rejects the claim that causation is transitive without explanation, perhaps because chance-raising is not transitive (e.g., 1998, pp. 118-9).][3]

But cases like that of example one, the late shooter, cannot be solved by Menzies' (1989) method (as he shows in his 1996), and I am unsure how Mellor would choose to deal with it. I think the best solution to the problem would be to adopt a more sensitive counterfactual conditional, so that C causes E only if the chance C gives E of being true at time t (when E actually is true) is greater than the chance of E at time t if ~C (Paul 1999).

However, since Mellor wants at least in part to reduce temporal relations to causal relations, and thus in general rejects assessments of probabilities of facts at particular times, he may not want to accept extant solutions to the problems.[4]

Preemption cases like these are the main sorts of tests applied to accounts of causation (especially in the last 10 years and increasingly so nowadays), in order to determine the viability of a metaphysics of causation. Standard regularity and counterfactual accounts fail these tests, and thus have been held to be inadequate. In some cases, suitably modified accounts can take their place, but in other cases the entire account must be reworked.

The importance of such cases is especially clear when we consider that in the case of the lucky shooter, if there were no second gunner (i.e., Sally is not around) Mellor's account would tell us that Fred's shot was the shot that killed Vicky, consistent with our common-sense judgments. But how could the presence or absence of a second gunner who in no way affects the causal process between Fred and Vicky change our verdict about whether Fred was a cause? The problem with preemption cases is that circumstances of the surroundings which have no ostensible connection with the intrinsic causal process under consideration nevertheless affect the verdict given by the theory of causation, which is strong evidence against the theory.

As I noted above, for Mellor, time is strongly linked to causation. Mellor argues that the fact that there is a temporal relation (i.e., later than) between C and E does not help us to use C to explain E, or show why C is a means to E. It is because there is a *causal* relation between C and E that we can do these things. Based on this Mellor argues that "if the general coincidence of causal and temporal order is to be more than a coincidence, it cannot be because time order fixes causal order. It must be the other way round [causal order fixes time order]" (1998, p. 108). If his theory of causation fails to specify necessary and sufficient conditions for causation, then it is picking out some other relation and reducing the temporal relation to this relation, whatever it may be. But if it is not the causal relation that Mellor has identified with his theory, then either Mellor cannot rely on claims about how the connotations of the causal relation (being a means, explaining) show us how and why we can reduce the temporal relation to the causal relation, or he cannot provide an illuminating account of what we have reduced the temporal order to (since it is not illuminating to claim that the temporal order reduces to the causal order, or call time "the causal dimension of spacetime" (1998, p. 115) without a substantive theory of what causation and hence the causal order is.)

The same point holds with other claims Mellor uses to drive his theory of time using his theory of causation. If time's link with causation tells us "how time differs from space, and why that difference makes time the dimension of change, gives it direction and makes it linear," then we need substantive information about the causal relation itself in order to make these claims interesting and evaluable.

There is one final option for Mellor that I have not yet presented. In *RTII* Mellor argues that facts are individuated by their causes and effects. If this view were workable, he could claim that he has no problem with preemption cases, since for him, there aren't any. If we individuate facts by their causes and effects, then we can maintain the requisite dependence of the fact caused by the preempting cause on that cause, since (necessarily) a different effect would have been caused if the preempted causal chain had run to completion. This would give Mellor an effective solution to the kinds of cases I discussed above. Unfortunately, as I discuss below, I think that the project of individuating facts by their causes and effects is a mistaken one, and that an artificial resolution of preemption cases only confirms the problems with adopting such a view.

2. FACTUAL CAUSATION AND CHANGE

The issue of the individuation of causes and effects arises in the context of Mellor's response to Davidson's Slingshot argument[5] against facts as causes and effects. To explain how causation keeps things unchanged and preserve his account of how causation drives changes of things, Mellor argues that there must be factual causation, i.e., one fact must cause another. If causation necessarily involved events or other particulars, unchanges would seem to create difficulties, since (according to Mellor at least) unchanges cannot be particular events, and thus when an unchange is caused (i.e., when we have stasis), there are no particulars that can be related and thus nothing that can be the basis for the causal relation. So Mellor holds that the causation of unchanges is fact causation: negative facts such as the fact that there is no cup of tea in front of me at t are among the causes of the fact that there is still no cup of tea in front of me at 2 seconds past t.

Mellor bases his account of change on the way these unchanges are handled (1998, p. 116-7): change involves a thing's having some (essential?) properties remain the same (unchanged) from t to t*, while other properties at t are replaced with incompatible properties at t*. The thing must have some unchanging properties in order for the thing to remain while it changes, i.e., in order to have a change in the thing rather than a complete replacement of one thing by another. The factual causation involved in unchanges also allows us to distinguish spatial variation from temporal variation, thus providing justification for calling the latter, and not the former, change. The distinction is based on the claim that causation cannot occur across spacelike intervals. Since according to Mellor's theory we need factual causation of stasis in a thing *together* with variation of (some of the) properties in a thing to have a change in a thing, variation across spacelike intervals cannot be change, since there cannot be any causation of stasis across these intervals. Since Mellor's account of change requires causation of unchanges (stasis), and since according to Mellor only fact causation can give us causation of stasis, Mellor's account of change rests heavily on his account of factual causation and thus of the facts which are the causes and effects.

The Slingshot argument is a proof that if "E because C" remains true both under arbitrary substitutions of fact-terms for necessarily equivalent fact-terms *and* under arbitrary substitutions of particular-terms for co-referring particular-terms, then everything causes everything. In *Real Time II*, Mellor rebuts the Slingshot argument against facts and factual causation by claiming that with respect to the sentence "E because C," singular terms which refer to the same particular(s) (as "C" and "E" do) cannot be substituted into "E because C" without change of truth value. In effect, the claim is that preserving the identity of particulars referred to is not sufficient to preserve the identity of the facts stated. This allows Mellor to deny the transparency of "E because C" with respect to co-referring expressions, the main premise of the Slingshot, but leaves him without an obvious replacement for a method of individuating facts. He fills this void with the claim that facts are individuated by their causes and effects: E=E* if E and E* have all the same causes and effects (1998, p. 104).[6] The claim has some *prima facie* plausibility: like the claim that events a and b must be the same event if *a* and *b* have the same causes and effects, at least for this-worldly events, the criterion seems to be correct. But there are problems with relying on such a claim in the context of providing a substantive theory of causation, especially a theory like Mellor's which relies on closest-world conditionals. If Mellor wishes to defend a theory of fact causation, he will need to find a different method of individuation. He needs *some* sort of individuation method for his facts; not only because this is a necessary part of a complete account of factual causation, but because his account of whether change occurs depends upon it. For if there is no account of how facts are individuated, then there is no account of how facts remain the same, and thus no account of the causation of stasis.[7]

The most serious problem with individuating facts by their causes and effects involves the use of the claim as part of one's case for a complete theory of factual causation. Mellor seems to want to provide such a picture: causes are facts that raise the chances of their effects, and we are to use information about how causes explain their effects and are means to their effects in order to develop the right characterization in order to, as Mellor puts it, discover what in the world causation really *is*. Certainly, if we are to have a complete story of factual causation, we must provide an account of the individuation of causes and effects as well as the necessary and sufficient conditions for causation. But individuating facts by their causes and effects compromises the project.[8]

This is because facts E and E* must be distinguished (i.e., we must have an individuation method) before the truth of the causal claim "E because C" is determinable. If the fact that E was not already distinct from the fact that E*, then the truth of "E because C" could depend on C's raising the chance of the fact that E*, since E* would be indistinguishable from E. In other words, E must be distinguished from E* in some way prior to the account of the truth of "E because C" in order to have a noncircular account. [Davidson, the original defender of the individuation of this-worldly events based on same causes and effects, abandoned his account (after Quine pointed out how it was circular) in favor of the claim that events are to be individuated by spacetime regions (Davidson (1985), p. 175).][9]

A second worry about the individuation of facts by causes and effects, even apart from the worry about circularity, involves the independent plausibility of the claim when investigated in the context of an analysis using counterfactuals. Mellor's theory depends on the evaluation of counterfactual claims about chances: for C to be a cause of E, the chance of E if C were false must be lower than the chance of E when C. In the closest possible world where C is false, the falsity of C must give E a chance p* which is lower than p, the chance of E when C. This means that we need to be able to identify E in close possible worlds as well as in our actual world in order to evaluate the chance of E in (appropriate) worlds where C is false (1995, p. 28). But the individuation criterion proposed seems to hamper this identification: intuitively, it seems that the fact that E, in nonactual but close possible worlds, could be the very same fact, otherwise caused, or could have different effects.

Example: we wish to see if the fact that Suzy drives her ball is a cause of the fact that she gets a hole in one. If the fact that Suzy drives her ball (C) raises the chance of the fact that she gets a hole in one (E), then the fact that Suzy drives her ball is a cause of the fact that she gets a hole in one. But to determine whether the fact that Suzy drives her ball raises the chance of the fact that she gets a hole in one we must move to the closest possible world (assume there is a closest world) where "Suzy drives her ball" is false, to see if in this world the chance of the fact that Suzy gets a hole in one is lower. Assume that it is. But it seems perfectly plausible that in this closest world, the very same fact that Suzy gets a hole in one could still occur, otherwise caused – perhaps she kicks the ball instead of drives it – or could have different effects. (Perhaps as the result of small differences between this possible world and the actual world, the fact that Suzy gets a hole in one causes the fact that there is a small dent in the cup in the possible world but not in the actual world.) In other words, there are certain intuitively clear cases where we want to be able to identify facts across possible worlds without having to assume that all their causes and effects must be the same. Indeed, if we do not identify facts independently of their causes and effects, then Mellor's account implausibly requires that for "E because C" to be true, the chance of E when C is false must be 0, since if C is a cause of E, then in the closest possible world where C is false, E must also be false, since by definition we cannot have E unless it is caused by C. Finally, as should be apparent, we cannot interpret the claim that facts are to be individuated by causes and effects as a this-worldly claim only, since then we are left without any means of evaluating the truth of the relevant counterfactuals, as there is no way to tell which other-worldly fact is or is not E.

3. BACKWARDS CAUSATION AND THE LINEARITY OF TIME

Mellor argues that his account of causation shows that there can be no such thing as backwards causation. This is needed for his project of reducing the temporal order to the causal order, for if we could have backwards causation his reduction would tell us that we could have times both earlier and later then other times – time could be circular, which Mellor denies (1998, ch. 11.1). Mellor also uses this argument to reject the possibility of backwards time travel, which depends on the possibility of backwards causation.

The argument in the final chapter of *RTII*, that there can be no backwards causation and hence that time must be linear, is based upon his view that the chances of an effect must be contingent, derived from contingent laws of nature. The problem with backwards causation boils down to the claim that the chances of causation in one direction must be logically independent of the chances of causation in the opposite direction. Mellor argues that if backwards causation were possible, the chances would not be logically independent, and so backwards causation is logically impossible.

I think Mellor's argument in *RTII* needs more development, since it seems that we can agree with Mellor that chances are contingent, but still have possible worlds where the laws of nature allow backwards causation and time travel. Example: imagine a world where the chances of effects are contingent: they are based on contingent laws of nature and the chances of effects are governed by these laws. In this world, we have the fact D, the fact that Dr. Who departs in his time machine in 2045 on his way back to 1945, wearing a scarf he found in a vintage shop. (The scarf is made in a style that was fashionable in the mid-twentieth century.) Dr. Who, wearing his scarf, arrives in 1945, intending to explore London (fact A), and the fact of his departing in his time machine with his scarf (D) has a 100% chance, shall we say, of causing A (the fact that Dr. Who, wearing his scarf, arrives in 1945, intending to explore London). Now, let's say that Dr. Who's intention to explore has a 10% chance of causing a craze for futuristic scarves like the one he wears during his explorations, resulting in one being available in a vintage clothing shop in 2045 which he buys and wears when he travels in his time machine.... In fact, his exploration causes the craze which results in his finding the scarf and wearing it, which then causes D, which in turn causes A ... and so on.[9]

Now, according to Mellor, the story I've just told is logically impossible, since A can only cause D if it is a property of A's circumstances that D has a chance d with A and a logically independent chance d* without A, and D can only cause A if it is a property of D's circumstances that A has a chance a with D and a logically independent chance a* without D. Furthermore, d and d* must be logically independent of a and a*: all combinations of individually possible a, a*, d, and d* must be possible (1998, p.133). Mellor claims that all combinations are not possible.

But it is not clear to me why Mellor thinks this. Let's fill in a few more details. Reviewing the information from above, we know that D has a 10% chance with A, and, let's say, a 5% chance without A. Aside from the obvious restrictions Mellor requires for causation to occur, these chances are logically independent: the chance of D with A could be any value d except 0, since D is a fact, and the chance d* of D without A need only be less than d, since D was in fact caused by A. Likewise with A: we specified earlier that A has a chance a of 100% with D, and let's say, a chance a* of 90% without D. So d=10%, d*=5%, a=100% and a*=90%.

All of these chances seem possible, and they all seem to be logically independent of each other in the appropriate way. So what can Mellor mean by claiming that such a story is logically contradictory? We might think it is bizarre or puzzling, as worlds that have time travel would certainly seem to be, but none of this amounts to a contradiction. I can think of only one way in which this story might seem contradictory. If we hold that A causes D and D causes A, and the chances are as specified above, then we can make the following claim: at 1945, the chance of D with A is 10%, and the chance of ~D with A is 90%. Now, at 2045, the chance of A with D is 100%, and the chance of ~A with D is 0. Further, the chance of A with ~D is 90%, and the chance of ~A with -D is 10%. This means that at 2045 the chance of A with A is 91%.

Perhaps this is the sort of result that Mellor is objecting to. But such results, while surprising, imply no contradiction. In a world where backwards causation is possible, the laws may be such that chances like these, where the chance of a fact at a time later than that fact is not 1 or 0, are possible.[10] So I do not know what more can be said in order to prove that backwards causation and hence time travel is logically impossible – in any case, I think the argument needs development if it is to be convincing.

CONCLUSION

That Mellor has made yet another important and creative contribution to the discussion about time is undeniable. I believe that his view is particularly interesting and best understood within the context of his other well-known works, *The Facts of Causation* and *Matters of Metaphysics*, and I encourage readers of *Real Time II* to read these as well in order to best appreciate Mellor's program. Although I would like to see the points I've discussed here resolved, I think that *Real Time II* is a significant achievement that, like its predecessor *Real Time*, will make a significant and long-lasting contribution to the debate about the nature of time.

NOTES

[1] Mellor has a complex view about the way his facts and facta are related to causation. Facta, things or truthmakers that are in the world, are what give facts their chances of being true (1995, 13.6). But causation is not a relation between facta: it is wrong to say that there is a relation in the world between facta that make causal statements true (or false) (1995, p. 165).

[2] Dorothy Edgington (1997) and Helen Beebee (unpublished) have also criticized Mellor's handling of preemption in *FOC* using similar cases.

[3] Paul Noordhof (1998, p. 863) suggests that Mellor could solve early preemption cases by using a modified version of Menzies' 1989 solution. Since Mellor holds that causation cannot involve unmediated action at a distance (1995, pp. 229-34), Noordhof suggests that Mellor's view be interpreted as: "a fact C is a cause of E if (i) it raises the chances of E against *some* background or other, G, and (ii) there is a chain of contiguous causes between C and E."

If Mellor adopted this view, he could say that Fred's shooting counted as a cause of Vicky's death because it raised the chance of Vicky's death against a background where Sally did not pull the trigger. (Whereas Sally's pulling the trigger is not a cause, since presumably there is no "chain of contiguous causes.")

I think this interpretation is unsatisfactory, and that Mellor should not adopt it. There are three reasons: (i) Noordhof himself notes that the account does not dovetail with Mellor's views about the connotations of causation (p. 863), (ii) the conditional is not informative, both because "cause" appears in the antecedent as well as the consequent, and because the definition of "chain" is unclear. (If by "chain" Nordhof merely means "sequence" then this is not a problem. But if by "chain" he means "sequence of facts which are causally related" then the notion, while probably correct, is unhelpful.) (iii) As I mentioned above, Mellor rejects the idea that causation must be transitive. But if causation need not be transitive, then the stipulation that there must be a contiguous chain of causes is unmotivated, except as an *ad hoc* attempt to rule out some preempted facts as causes. On the assumption that causation is transitive, the requirement that there is a contiguous chain of facts that are causally linked from C to E helps us identify C as a cause of E and rule out non-causes of E. But if causation need not be transitive, then the fact that there is a chain of causes from C to E does not help us to establish C as a cause of E.

[4] See Paul (1998a) for a discussion of late preemption cases that shows there need be no missing events (and by extension, facts) in the preempted causal chain.

[5] Donald Davidson (1980), Essay 7, "Causal Relations," pp. 152-153.

[6] There is room for some confusion here. Mellor distinguishes between facts and facta: the latter are entities in the world which can be the truthmakers for facts. In *Real Time II*, Mellor uses "fact" to refer to facts, but *also* to refer to facta, but only "when it matters," p. 25. It was not always clear to me when it mattered. Presumably, Mellor is talking about facts rather than facta when proposing his individuation criterion for facts, since he was responding to Davidson's slingshot argument, which is an argument directed towards facts as Mellor defines them. On this assumption, I have left the problem of individuation as it could pertain to facta aside.

[7] An individuation method might also aid in resolving preemption cases, and in bolstering the case against the claim that facts (in non-preemption cases) which do not raise the probability of effects are not causes.

[8] Advocates of event causation generally recognize that they must have an account of the individuation of events to go along with a complete account of causation, and most try to provide a method of individuation. Many of these theories of events have serious problems, which translate into serious problems with the theory of causation they are supposed to support.

[9] I'm indebted to David Lewis here.

[10] Lewis (1986) acknowledges this sort of case as an exception.

REFERENCES

Beebee, H., unpublished manuscript.

Davidson, D. (1980), *Essays on Actions and Events* (Oxford: Clarendon Press).

Davidson, D. (1985), "Reply to Quine on Events," in LePore and McLaughlin (eds.), *Actions and Events: Perspectives on the Philosophy of Donald Davidson* (Oxford: Blackwell): 172-175.

Edgington, D. (1997), "Mellor on Chance and Causation," *British Journal of the Philosophy of Science* 48: 411-433.
Lewis, D. (1986), "A Subjectivist's Guide to Objective Chance," reprinted with postscript in his *Philosophical Papers,* Vol. II (New York: Oxford): 83-132.
Mellor, D.H. (1981), *Real Time* (Cambridge: Cambridge UP).
Mellor, D.H. (1991), *Matters of Metaphysics* (Cambridge: Cambridge UP).
Mellor, D.H. (1995), *The Facts of Causation* (*FOC*) (London: Routledge).
Mellor, D.H. (1998), *Real Time II* (*RTII*) (London: Routledge).
Menzies, P. (1989), "Probabilistic Causation and Causal Processes: A Critique of Lewis," *Philosophy of Science* 56: 642-663.
Menzies, P. (1996), "Probabilistic Causation and the Pre-emption Problem," *Mind* 105: 85-117.
Noordhof, P. (1998), "Causation, Probability, and Chance," *Mind* 107: 855-875.
Paul, L.A. (1998), "Problems with Late Preemption," *Analysis* 58 (1): 48-53.
Paul, L.A. (1999), "Keeping Track of the Time," *Analysis* 58 (3): 191-198.

7

TIME, CONSCIOUSNESS AND THE KNOWLEDGE ARGUMENT

John Perry
Stanford University

INTRODUCTION

Hugh Mellor's *Real Time II* is an excellent book, for the following reasons. First it deals with an important and difficult philosophical problem, the passage, or apparent passage, of time. Second, Mellor is basically right. Mellor is a B-theorist, in the sense that he thinks that the B-facts, the facts that are not relative to a time, provide all the truthmakers we need. I was convinced that this must be right by reading D.C. William's "The Myth of Passage" a long time ago, when I was in graduate school. But although I knew which side I was on, this conviction was based more on seeing problems for the A-theory, and thinking that the A-arguments weren't very convincing, than on understanding how the B-theory could account for everything. And this is the third thing I like about Mellor's book. As I read it, I really felt, for the first time, that I understood, or at least began to understand, how the B-theory not only must be right, but could be right.

In this paper, I will argue that at the heart of the A-position is an argument analogous to Frank Jackson's knowledge argument and that Mellor'sapproach in *Real Time II* can be extended to deal with Jackson's argument.[1]

TIME AND THE KNOWLEDGE ARGUMENT

Jackson's famous argument goes as follows. There is a young woman raised in a Black and White Room, and never allowed to have any color experiences. She knows all the physical facts, for she is quite bright and reads all the books there are to read, and from them learns all the physical facts. But when she is let out of the room, she learns what it is like to see red. So she learns something, a fact, that she did not know. So there is a

L. N. Oaklander (ed.), The Importance of Time, 81–93.

fact she did not know. It isn't a physical fact, for she knew all of them, so it is a non-physical fact, and so there are non-physical facts.

Suppose now that there is a young man named Larry, who is raised in a room in the Claremont Hotel with no up-to-date calendars. That is, he is never allowed to know what day it is, or even what year it is. This young man has a passionate interest in the Pacific Division of the APA, and is allowed to read all of the APA publications there ever have been, including programs of all of the past APA's and also future APA's for the next twenty years. (Let's pretend that Anita Silvers is part of this experiment, and that, motivated by a terrific deal on printing programs, she has decided on the schedule of talks for the next twenty years. She will rely on her vast knowledge of the membership of the Pacific APA to make sure things turn out right, making deft suggestions to the key committee members at the right times.) So Larry knows all the B-facts there are to know about events at the APA meetings: which talks come before which talks, which sessions are concurrent, and so forth. He knows, for example that in 1999 on April 1 there was or will be a meeting about the views of Hugh Mellor. He knows also

(1) The 1999 Presidential Address is on April 2, 1999.

Every year or so he can tell, from the small window in his room at the Claremont Hotel, that the Pacific APA meetings are starting, because he can see the sign out in front that says, "Welcome Pacific Division, APA." But he doesn't know what year it is. Larry waits patiently for his release, but he is proud to be part of a philosophical experiment. As a matter of fact, Larry has just been released, stepped out of the room in which he was kept, and joined us here at the symposium. Larry peeked at the Palm Pilot of the person sitting next to him - he first up to date calendar he has seen - and realized:

(2) Today is April 1, 1999.

and from that and what he already knew already, he inferred

(3) The 1999 Pacific Division APA Presidential Address is tomorrow.

(3) is new knowledge for Larry. (3) expresses something that Larry did not know before his release, while (1) expresses something he did know before his release. So (1) and (3) seem to be different bits of knowledge. And they both seem to be APA facts in the sense above, facts about when the APA sessions are held. But (3) cannot be identified with (1), since Larry knew

(1) without knowing (3). The same goes for any other of the APA B-facts Larry knew. So there are not just APA B-facts. There are non-B APA facts as well, and these are what we all call A-facts.

SPACE AND KNOWLEDGE ARGUMENT

In *Real Time II* Mellor often defuses arguments about time by looking at the analogues for space. Space seems to be inherently less mysterious than time, and so we are less willing to accept arguments that lead to mysterious conclusions. This is a well-known strategy in teaching philosophy. Students will accept forms of argument if they are presented as having some fairly profound conclusion, say that people don't really survive from moment to moment. But they will see through the forms of argument very quickly if they are used to show something relatively trivial, say that the ham sandwich they are about to eat really doesn't belong to them, since it really isn't the one they paid for at the counter (it was at the counter, and this one is on their plate, and so they have different properties, so they are not the same).

Can we construct a knowledge argument for space? Certainly. Put Larry in a windowless hut across from Little America, just off Interstate 80 in western Wyoming (Little America is a gas station with a restaurant and souvenir shop. It has more gas pumps than anyplace in the world.) Give him an Interstate Road Map to memorize. Larry may know all the B-type Interstate 80-facts–the order of states, cities, towns and villages as one progresses east to west or west to east along Interstate 80, from Berkeley through Reno, Salt Lake City, Little America, Cheyenne, Lincoln, and so on. But he isn't allowed to look out of his hut so he doesn't know where he is. Eventually let him out. He sees all the gas pumps, realizes he is in Little America, and immediately knows a number of facts that seem to be Interstate 80 facts, although they are not B-facts. He already knew the B-fact

(4) Salt Lake City is west of Little America.

Now he learns,

(5) Little America is here,

and infers

(6) Salt Lake city is west of here.

And so on for many other things. Since he knew all of the 80 B-facts before he was let out, but didn't know these facts, these are clearly not B-facts, so there are Interstate 80 A-facts as well as Interstate 80 B-facts.

BELIEFS AND FACTS

I think the key move in the knowledge argument is the move from

(7) Larry (or Mary) has a new bit of knowledge

which seems to me to be true, to

(8) There is an X-fact that Larry (or Mary) did not know

where X is the type of fact at issue, viz., about the mind in Mary's case, about the APA in our first Larry case, and about Interstate 80 in our second. I think this step is a fallacy, which rests on a mistake about the structure of knowledge.

Mellor's conceptual framework in *Real Time II* provides an analysis of what is wrong with this argument. Mellor is not an across the board enemy of A things. He allows A-representations, but not A-facts. He allows A-perceptions, A-thoughts, A-utterances, A-propositions and A-beliefs. But he does not have A-facts or *truthmakers*. Once you have all of the B-facts, you have all of the facts. You have all of the truthmakers for the A things as well as the B things. An inventory of the B-facts of a world is a complete inventory, nothing is left out.

Mellor thinks of A-beliefs as beliefs in A-propositions. We can say that Larry did acquire new beliefs, beliefs in A-propositions, that he did not have before. And these beliefs were true. So he acquired bits of knowledge he did not have before. But it does not follow from this that there are facts he didn't know before, only that there are true A-propositions that he didn't believe before. So (8) simply does not follow from (7).[2]

Think of a possible world as telling us, for each time, what happens. Call these the B-facts of the world. Think of a proposition as a set of such possible worlds. Think of a B-belief as a relation that an agent has to such a proposition at a time. If the proposition is true, call this B-knowledge. Then the inference from 7B to 8B seems valid:

(7B) Larry has a new bit of B-knowledge
(8B) There is a B-fact that Larry (or Mary) did not know

Let an agent and a time and a place, where the agent is at the place at the time, be an *occasion* of belief. Let an A-proposition be a function from occasions to B-propositions. A belief in an A-proposition is a true belief, if the B-proposition that is the value of the A-proposition at the occasion of belief is true.

For example, Larry's A-belief, that he expresses with

(6) Salt Lake city is west of here,

is a belief in an A-proposition. This proposition is a function that will yield the B-proposition that Salt Lake City is west of *p* at an occurrence *a, t, p*. The truthmakers for Larry's belief consist of

(i) The facts of the occurrence of the belief, that it is a belief of Larry's at *t* at Little America (where *t* is the time his utterance);
(ii) The facts about the subject matter of the belief, that at t Salt Lake City is west of Little America.

Both of these truthmakers are perfectly good B-facts. There is no need for A-facts. So Larry has a new bit of knowledge, but this knowledge doesn't consist of knowing a new fact. The knowledge argument gives us no reason to suppose that there are A-facts in addition to B-facts.

Now let's turn to Larry, who until a few minutes ago was locked in the Claremont without an up to date calendar. Larry's beliefs change as he emerges and sees the date on the Palm Pilot next to him. He comes to believe something he expresses with

(2) Today is April 1, 1999,

and this leads to many more new beliefs.

Larry's new belief is a belief in an A-proposition, a function from occurrences of belief *a,t,p* to the B-proposition that at *p, t* occurs during April 1, 1999. This would exclude worlds in which Larry emerges from the room just has he does, but the room is in Tokyo, where the present moment is a part of April 2, 1999, not April 1st. It excludes worlds in which Larry comes out on April 4, 2003, and sits next to someone with a defective Palm Pilot, which leads him to say (2). But in fact Larry is here in the Claremont, and emerged today, so his belief is true. The truthmakers are:

(i) the fact about when and where his belief occurs, namely, right at this moment in the Claremont,

and

(ii) the fact that in California, where the Claremont is, this moment is a part of April 1, 1999,

which are both perfectly good B-facts.

In both cases, then, when Larry acquires a new bit of knowledge, he does so by acquiring a new A-belief which is true. But the truth of the A-belief does not require any A truthmakers or facts.

THE CLASSIC KNOWLEDGE ARGUMENT

If we follow Mellor, and break the move from (7) to (8), we can diagnose what is wrong with the knowledge argument. Mary emerges from the Black and White room and sees a fireplug. She knows, as a matter of general knowledge picked up in the Black and White Room, that fireplugs are red. She attends to the experience she has when she looks at the fireplug, the experience of seeing red, and forms a belief she might express with

(9) This is what it is like to see red,

or perhaps,

(10) This quale is the what-it-is-like aspect of the experience of seeing red,

or perhaps

(11) This quale is the seeing-red quale.

Mary has a belief she didn't have before, that she expresses in one of these ways.

ANTECEDENT PHYSICALISM

Now let me digress for a moment, and quickly explain a perspective that I call "antecedent physicalism." The word "antecedent" is a nod to Hume, in the

Dialogues on Natural Religion, where he observes that although all the pain and suffering we find in the world are not what we would expect from an all-perfect God, they do not logically contradict the existence of such a being. Hence, if we are *antecedently convinced* of such a God, these phenomena will not disprove it.

Similarly, qualia, the what-it-is-like aspects of experiences, are not something that strike us as all that physical. But it seems to be that if we are antecedently convinced of physicalism, they do not give us a good reason to abandon it.

The antecedent physicalist (me), is not skeptical about qualia, but quite enamored of them. Consider the taste of a warm Mrs. Fields chocolate chip cookies as you chew it, trying hard not to swallow so the qualia will last as long as possible. Or the taste of a handful of zinc-covered nails, that you accidentally put in your mouth while working on the roof, forgetting that the mouth is not a good storage spot for zinc covered nails, although it works well for most other kinds.

Since I am an antecedent physicalist, I must regard these experiences as physical states and processes, brain-events, as it were. And I also think they have only physical properties. So the aspect of my experience, of my brain state, that I am aware of and enjoy as I eat the cookie, is a physical aspect of my brain state.

This view has often thought to be absurd. Leibniz asks us to imagine a brain as big as a factory. Walk into it. Could you see the experience of tasting a chocolate chip cookie? A.C. Ewing asks the physicalist to pick up a hot coal from the barbecue. Carefully attend to the searing, unendurable pain that rises in your consciousness. Does that seem like a brain state?

The fallacy of these arguments is that there is no reason to suppose that what-it-is-like to see a brain state, is anything like what-it-is-like to be in that brain state. The knowledge argument, I think (and Chalmer's Zombie argument too, by the way), is simply a fancy version of the Leibniz-Ewing argument, a fancy version that allows us to see the structural relation between this argument and Passage of Time arguments.

So, metaphysically, the antecedent physicalist thinks that experiences are brain events, and qualia are real physical properties of those brain events. There are lots of sentient beings, in the sense of having brain states such that it is like something to have them. The antecedent physicalist believes sentience is a central and important part of the architecture of the beings that have it. The basic idea is very old-fashioned. A basic part of being an animal is to avoid pain and seek pleasure.

We humans, and no doubt many other kinds of animals as well, are not only sentient, we are aware of our sentient states. That is, we not only have brain states it is like something to be in, we note such states, think about them, classify them, notice the kinds of situations in which they occur, and adopt elaborate plans to avoid being in them or promote being in them. I might base my whole day, for example, on walking past a Mrs. Fields shop so I can get a cookie. I may brush and floss every day for years to avoid having a cavity that needs to be drilled and filled. I make these plans because I know what it is like to eat Mrs. Fields cookie, and I know what it is like to have one's tooth drilled, and I think I know what sorts of situations lead to, and prevent, experiences of these sorts.

This inner awareness we have our own what-it-is-like states has some analogies with perception, and some disanalogies. There is not as far as we know some kind of organ, like the eye or the ear, involved in awareness of our experiences. Nor is there, as far as we know, a transmission through a medium of some sort of disturbance, as with hearing and seeing. But there are clear analogies with perception of things that are close by. We can only be aware of experiences when we have them. Later on we can remember them; beforehand we can anticipate them, but we can only be aware of them when they occur. And, like the things we are looking at or listening to or smelling, we can direct our attention and focus to some of the things we are aware of. I can focus on my experience of red, or my headache, or the minor pain in my finger.

And, as with things we perceive, the natural way to talk about what I am aware of, when I attend to it and focus on various aspects of it, are demonstratives: "this feeling," "this emotion," "this ache," and the like.

BACK TO MARY

Armed now with Mellor's break between objects of knowledge and facts, and the perspective of antecedent physicalism, let's return to Mary in the Black and White Room. The natural way for Mary to express her new bit of knowledge, when she leaves the Black and White Room is,

(9) This qualia is what its like to see red.

How should we think of this use of "this qualia?" "This qualia," in the use Mary is making of it, will refer, on a given occasion of awareness, to the qualia to which the agent is attending. So, in Mary's case, it will refer to the what-it-is-like aspect of her experience of seeing red. But, in other cases, it will refer to other qualia. Suppose Mary is confused, and a bit later, while looking at something that is lavender, and attending to the lavender quale, utters (9) again. In this case, "this qualia" will refer to the quale she is attending to, the quale that occurs when people look at something lavender, and so she will have said and thought something that is false.

From our physicalist point of view, we want to think that all the facts there are, are the physical facts. They play the role of the B-facts in this dialectic. And we think that the physical facts *are* B-facts; that is, physical events happening at times. So we want the truthmakers for Mary's beliefs to be physical facts. The knowledge argument challenges this, but using lessons from Mellor we can see what to say.

Let us say that physical B-worlds are B-worlds in which only physical things happen. Physical B-propositions are sets of such worlds. And physical A-propositions are functions from episodes of directed inner awareness to physical B-propositions. An episode of directed inner awareness will involve an agent, a time, and a quale, such that the agent is having and attending to the quale at the time.

Let us label the quale which a person with normal vision has when he sees something red in normal circumstances Q_r. Mary is attending to Q_r, in the original case. So, what we want to say about Mary is:

(A) The object of her belief is the physical A-proposition expressed with "This qualia is what it is like to see red."

(B) This physical A-proposition is a function *f* from occasions of directed inner awareness *c* to physical B-propositions *P*:

$f(c) = P$, where for each world *w* in *P*, in *w* the quale that the agent of *c* is attending to at the time of *c* is what it is like in *w* to see red.

(C) The truthmakers for Mary's belief are the B-facts:

(i) Mary inner attends to Q_r
(ii) Q_r is what it is like to see red

both of which are physical B-facts.

On this account, there is nothing about Mary's case that counts against physicalism.

THE MYTH OF OBJECTIVE KNOWLEDGE

The knowledge arguments, in all their applications, move from a new bit of knowledge, to a newly known fact. This is the step from (7) to (8):

(7) Larry (or Mary) has a new bit of knowledge
(8) There is an X-fact that Larry (or Mary) did not know

The strategy we have been looking at for blocking this step is to allow two kinds of belief, and hence two kinds of knowledge. For B-belief or knowledge, belief is a relation to a B-proposition. A B-proposition is true or false independently of who believes and when or where; it is objective. In contrast, we have A-beliefs, which are beliefs in A-propositions, which are not true independently of who believes them and when and where.

Now I'd like to explore a somewhat different option, a somewhat different way of looking at this. This option says that there are not really two kinds of belief. There is really just one kind, with all sorts of variations. Each episode of belief involves a believer in a belief state at a time, a way of believing, and a B-proposition believed. What differs is the extent to which the same ways of believing yield the same proposition believed, given changes in the facts about the episode. Being in the sort of belief state one expresses with "It is raining here now" constitutes belief in quite different propositions, depending on when and where the belief takes place. Being in the sort of belief state one expresses with "It is raining in Berkeley, California, now" constitutes belief in different propositions depending on the time of the episode, but the place does not matter. Being in the sort of belief state one expresses with "2 + 2 = 4" constitutes belief in the same proposition, no matter who is in the state, when, and where. So on this view, we characterize the belief-state with something like what Mellor calls A-propositions, but we don't think of the entities as *propositions*, but as abstract objects that characterize an important property of belief states, by characterizing them as relations between the relevant facts about the belief episode, the context, and the proposition believed.

Let's start with Larry's "A-belief" of twenty minutes ago, that he expressed with "Today is April 1, 1999." On our treatment of this belief, there are three basic elements:

(i) The circumstance of the belief: the agent, time and place where the belief occurs (Larry, a time **t** about twenty minutes ago; this room in the Claremont);
(ii) The nature of the belief; (It is the kind of belief expressible by "Today is now April 1, 1999");
(iii) The proposition Larry believes in virtue of having that sort of belief in those circumstances, the B-proposition that in Berkeley **t** occurs on April 1, 1999.

If we fix the nature of a belief, we establish a *relation* between the other two factors, the circumstance of the belief, and the proposition believed. That is, given we are dealing with a type ii) belief, what we fill in for i) constrains what we can fill in for ii) (and vice versa).

Determining the circumstances of the belief, who has it, where, and when, determines what the subject matter has to be like for the belief to be true, that is, which proposition is believed. For example, once we have fixed the nature of the belief as being of the same type that Larry had, the type one expresses with (3),

(3) The 1999 Pacific Division APA Presidential Address is tomorrow

the following pairs of "belief context" and "belief content" will be appropriately related:

First pair:

(i) Yoshi, Tokyo, and the time **t'** 24 hours before **t;**
(iii) That the 1999 Pacific Division APA Presidential Address is/was/will be April 2, 1999.

That is, twenty four hours ago, by being in the same state Larry is in now, Yoshi would have believed the same B-proposition is Larry now does. Of course, if he were in that state as the same time as Larry is - that is, right now - he would not believe the same thing as Larry, since it is already April 2 in Japan. And if Larry had been in that state at **t'**, he would have believed something false, that the Presidential Address was April 1, since, in Berkeley, **t'** occurred during March 31.

Second pair:

(i) Malcolm, London, and the time **t'** 4 hours before **t**
(iii) That the 1999 Pacific Division APA is/was/will be April 2, 1999.

And so forth...

What I want to suggest, then, is that we can characterize the contribution of the subject's brain, to the truth or falsity of the belief, with a relation between (i) B-facts and (iii) B-propositions, what we might call generally, *facts about context of representation facts* and *the proposition believed.*

What Mellor calls A-propositions are better thought of, I think, as abstract object characterizing these belief types.

Compare now

(4) Salt Lake City is west of Little America.
(6) Salt Lake city is west of here.

Whether you think of (4) and (6) as English sentences, or as the belief states that might be expressed with such English sentences, they differ dramatically in the relationships they impose. An occurrence, belief or utterance, of (4) will be true no matter where it is spoken or thought, any time after the construction of Little America and before some cataclysm destroys one or the other or moves the surface of the U.S. around quite a bit. An occurrence of (6) will be true if it is spoken or thought at some place *p* and time *t*, such that it is a fact that *p* is east of Salt Lake City at *t*. The difference isn't that what is believed in the case of (6) is much more dynamic and fragile than what is believed in (4). It is rather than the sentence (or type of belief) in (6) has a much more fragile relationship with the non-fragile truth that Salt Lake City is west of Little America. It only expresses that truth from some places. So, in a way, one needs to know more to assert (4) than (6), one needs to know the facts of one's context of utterance or belief, which is what Larry learned when he stepped out of the hut.

This relational perspective is more or less that recommended in my 1983 book with Jon Barwise, *Situations and Attitudes.* Since that time I have developed an account I call the reflexive/referential account, and applied it to a number of philosophical issues, including the knowledge argument. The underlying picture of belief states as classifiable by relations between facts about the episode and states of affairs involving the subject matter remains intact, however.

NOTES

[1] See Robert Coburn (1990) for Jackson's argument. See A. C. Ewing (1962) for a discussion of A and B dichotomies with respect to time, space, and possible worlds.

[2] The way Mellor thinks of this isn't quite the same as I do; I say a little bit about this below, and quite a bit in John Perry (1997) and (1993). The differences between us have to do with the philosophies of language and propositional attitudes, however, rather than with than the issues about time and consciousness that concern us here. The key point is that Mellor's set-up allows us to block the step from (7) to (8).

REFERENCES

Barwise, J. and Perry, J. (1983), *Situations and Attitudes* (Cambridge, MA: Bradford-MIT Press).

Chalmers, D. (1996), *The Conscious Mind* (New York: Oxford University Press).

Ewing, A.C. (1962), *The Fundamental Questions of Philosophy* (New York: Collier Books).

Coburn, R. C. (1990), *The Strangeness of the Ordinary: Problems and Issues in Contemporary Metaphysics* (Savage, MD: Rowman and Littlefield).

Jackson, F. (1986), "What Mary Didn't Know," *Journal of Philosophy,* LXXXIII: 291-295.

Mellor, H. (1998), *Real Time II* (London: Routledge).

Perry, J. (1993), *The Problem of the Essential Indexical and Other Essays* (New York: Oxford University Press).

Perry, J. (1997), "Indexicals and Demonstratives," in Robert Hale and Crispin Wright (eds.), *Companion to the Philosophy of Language* (Oxford: Blackwells Publishers Ltd.): 586-612.

8
REAL TIME II: REPLIES TO HINCHLIFF, PAUL AND PERRY

D. H. Mellor
University of Cambridge

I'd like to start by thanking the organisers of this symposium, and especially Nathan Oaklander, to whom I and all other B-theorists are also indebted for his most effective championing of our cause over many years. I'm also very grateful to all my three commentators, both for their kind remarks about my book, and for the time and trouble they've taken to produce such important and incisive comments. I hope therefore that they will forgive me if, in order to leave as much time as possible for general discussion, I reply only briefly, and only to their main criticisms, leaving unstated, but not unappreciated, our points of agreement.

MARK HINCHLIFF

I start with Mark Hinchliff. As an unregenerate presentist, Mark offers one main objection to my argument against A-truthmakers. We both agree that if the A-fact that *e* is past is what makes tokens of the A-proposition that *e* is past true, then tokens of that proposition produced before *e*, when they are false, must become true after *e*, when the fact that *e* is past comes into existence. This implies for example that an undertaker who mistakenly buries me alive will have his then false token belief – that my life is past – made true later by the fatal consequence of his mistake. I deny this – and so I hope will the jury when the undertaker is subsequently tried for acting with criminal negligence on a culpably false belief! More seriously, and more generally, to suppose as Mark does – since he has to – that my undertaker's false token belief is made posthumously true by my death is to take the truth values of indexical tokens to be fixed not by when or where they occur, or by who produces them, but by when, where or by whom their truth values are assessed.

L. N. Oaklander (ed.), The Importance of Time, 95–102.

With all due respect to Mark, this view still seems to me obviously absurd. If you phone me to say "I have a headache: it's so hot here in Mexico," I will not reject that token statement of yours as false because *I* have no headache and am not in Mexico. And it seems no less obvious to me that premature and therefore false tokens of "*e* is past" do *not* become true when – and just because – *e is* now past. Yet unless they do, that A-fact cannot be what gives those tokens their truth values. And if it is not, then we have no reason to postulate it or any other A-truthmaker for A-beliefs and A-utterances.

Mark's other objection is to my new non-temporal-parts B-theory of change. Suppose a thing *a* has a real changeable property *F* (like a temperature) at one time *t* and also at a later time *t'*. I say the only truthmakers here have the same two constituents – *a* and *F* – with *t* and *t'* being not *constituents* of these two *a*-is-*F* facts (as I shall call them) but merely their temporal locations. And then I say that for *a* to *change* from being *F* at *t* to not being *F* at *t'* is for *a* to *be* at *t'* – by having some other property there – but for there to be no *a*-is-*F*-fact at *t'*.

To this account of change Mark objects that it makes *F* what he calls a "relativised property," i.e. one which a thing can only have relative to something else, in this case a time, thus making the B-proposition that *a* is *F* a relativised proposition, i.e. one whose truth value is relative to a time.

Well, I agree that when the B-*sentence* "*a* is *F*" doesn't just mean that *a* is *always F*, its truth value is indeed relative to a time. But in that case I don't think this *B*-sentence – unlike the *A*-sentence "*a* is *now F*" – expresses a proposition at all. The only B-propositions here that need their own truthmakers are the propositions that *a* is *F* at *t*, *a* is *F* at *t'*, and so on, propositions which I say are made true (or false) by there being (or not being) *a*-is-*F* facts located at those times.

Nor does my theory make *F* a relativised property in Mark's sense. For I deny that *a*'s having this property (when it does) is relative to anything, since none of my *a*-is-*F* facts depends for its *existence* on anything other than spacetime. Different instances of *F* do of course need different spacetime *locations*; but then so do different qualitatively identical electrons, and no one infers from this that being an electron is a relativised property. And if it isn't, then nor is *F*.

The fact that different B-truthmakers, like my *a*-is-*F*-facts, are *located* at different times doesn't mean that only at those times is it true to say – tenselessly – that they *exist*, i.e. that their *existence* is relative to those times. It isn't – and it doesn't need to be in order for these B-facts to give the *A*-proposition, that *a* is *now F*, all its temporally variable truth values: since all that requires is that *a*-is-*F facts* be located at all and only the B-times at which this A-proposition is true.

Finally, in a note added after the session, Mark asks why my truthmaker for "the candle is bent at *t*," namely a candle-is-bent fact located at *t*, doesn't also make true "the candle is bent." The answer is, as I have said, that *B*-sentences of this form "*a* is *F*" (as opposed to "*a* is always *F*") have no truthmakers because they have no truth values, since they express no propositions.

So I still see nothing wrong with my new B-theory of change. I continue to think that it squares a very old circle by letting *a* itself, and not just temporal parts of *a*, have a monadic property *F* at some B-times and not at others, without contradiction.

LAURIE PAUL

Turning now to Laurie Paul's comments, she is of course right to say that much of my theory of time depends on my theory of causation. The dependence goes the other way too, since I give other grounds for rejecting the A-theory of time and hence all theories of causation – like John Mackie's (1974, ch. 7) – which assume it. However, as Laurie isn't an A-theorist, we needn't discuss A-theories of causation, and so I can turn directly to her main objections to my B-theory.

First, there are her apparent counter-examples to my claim that causes must raise the chances of their effects. I argue that causes must do this if causation is to keep enough of its consequences to be worth calling "causation," and in particular to make causes *ipso facto* explain, be evidence for, and means to, their effects. (This incidentally – since Laurie raises the question in a footnote – is why I deny that causation is transitive, because I think transitivity fails for all these connotations, and I see no independent reason to accept it. It is after all what generates all those incredible cases of storms being caused by the flapping of butterfly wings.) I do however accept that causation is *dense* – i.e. that there is no unmediated causation at a distance – for reasons which I give in *The Facts of Causation* though not in *Real Time II*; and this, as Laurie notes that Peter Menzies (1989) has shown, enables us to reject her so-called "early pre-emption" counter-examples.

However, as she says, denseness will not dispose of her *late* pre-emption examples, and these I would in fact deal with much as she would. For if – in her example – I say that Vicky dies because Fred shoots her, my B-statement of the effect, "Vicky dies," must, if Vicky is mortal, be elliptical. I must mean something like "Vicky dies when she does" and this will indeed have a higher chance of being true for *some* time interval that spans her death if Fred shoots her when he does than if he doesn't. And then, since a fact about Fred's shot (namely when it occurs) causes a fact about Vicky's death (namely when *it* occurs), it follows on my theory of causation that his shot causes her death, or at least affects it, by causing it to occur earlier than it otherwise would.[1]

Yet Laurie says, puzzlingly, that I may not want to accept solutions of this kind because I "want to reduce temporal relations to causal relations and thus in general reject assessment of probabilities of facts at particular times." But I do no such thing. For a start, I don't think that temporal relations *do* reduce to causal ones, if only because quantitative relations like *being ten minutes later than* don't. All causation does for me, as Laurie notes, is distinguish time from space, give it its direction and make it the dimension of change. That I admit is quite a lot; but it isn't everything, and in particular, it isn't anything metric.

Secondly, since we can use many other causes and effects to identify the spacetime region in which Fred fires (or fails to fire) his shot, I can and do say that Vicky's dying when she does has a chance in that region (the chance that is a property of facts located there) which can easily differ from its chances in other regions, such as the much greater chance that her dying when she does has at the time and place at which Fred's bullet enters her body. So I can easily accept solutions to late pre-emption problems of the kind that Laurie proposes.

Next, Laurie objects to my use of causation to individuate the facts which I say are causes and effects. To this I reply first that a criterion of identity is no more needed for all facts or all events than for all the concrete particulars, ranging from viruses through people and mountains to galaxies, whose lack of clear identity criteria leads nobody to doubt their existence. The fact is that entities don't *need* identity *criteria* – they just need *identity*, which comes free. Calling facts that we can use to individuate entities of a given kind (like facts about people's spacetime locations) "criteria of identity" doesn't make those facts essential to those entities. Laurie's world-line, useful as it is for distinguishing her from other people, is not what makes her the person she is.

I hijacked Davidson's (1969) causal criterion of identity for causes and effects to show *ad hominem* that it works just as well for my facts as it does for his events, and also to show what's wrong with his slingshot argument against facts (Davidson 1967). But you don't really need an identity criterion for facts to reject Davidson's slingshot: you need only reject one or both of its two assumptions, simply because neither is obviously true, and they do have obviously false consequences.

Actually I did more than that in *The Facts of Causation*, since the opacity of probability statements let me use my chance-raising condition on causation to show when and why the transparency assumption of Davidson's slingshot fails. But that is a bonus, and irrelevant to my reply to Laurie: namely, that there is nothing wrong with theories of causation, like Davidson's and mine, which postulate causal relata with causal criteria of identity. That is no more viciously circular than scientific theories postulating entities, like fundamental particles, with criteria of identity (if any) that only they can supply. Theories and criteria of identity for the entities they postulate can, and often do, come as package deals, and are none the worse for that.

Nor does my causal criterion entail, as Laurie thinks, that a given fact could not have different causes or effects – any more than a world-line criterion of her identity implies that she had to be here today. All my criterion says is that a fact D is identical with a fact D' if D and D' *actually* have all the same causes and effects: it says nothing about how different that fact's causes or effects *could* be without making it a different fact. So in particular, it doesn't stop an effect E of a cause C being the very same fact in those nearby non-C worlds where I say not-C must give E a lower chance than C gives it in our world. In other words, mine is a criterion only of *actual*, not of *counterfactual*, identity; and again, none the worse for that. No one thinks our metaphysics of particulars is incomplete because we cannot say in general how different a galaxy, a mountain, a person or a virus could be without losing its identity; and what's sauce for those particular geese must surely be sauce and for my factual ganders.[2]

Finally, backward causation and backward time travel, which I say cannot happen because they entail causal loops that I argue are impossible. We can of course describe worlds with such loops, as Laurie does, just as we can describe worlds with barbers who shave all and only those who don't shave themselves. That doesn't show that these imaginary worlds are *possible*, and I say they're not. Take the simplest case, in which two facts C and E cause each other. This on my theory requires E to have chances with and without C, which are properties of facts where C is, and C to have chances with and without E, which are properties of facts where E is. So here we have four chances: actual and counterfactual chances of E and of C, located in two different parts of spacetime. I therefore make what I shall now call the "Humean" assumption that none of these chances can entail *anything* about the others, an assumption I stated properly on p. 132 of *Real Time II* but then for brevity unfortunately expressed by calling these chances "logically independent."

But that description of them is too weak, as Laurie points out, since even in a causal loop no *one* value of any of my four chances need be entailed by the other three, and in that sense they can be logically independent. Many combinations of them are indeed compatible with a causal loop. But not *all* combinations of them, which is what I meant by calling them logically independent and will now mean by calling them *Humeanly* independent. And from this assumption, that all individually possible values of any effect's actual and counterfactual chances must be compatible with each other and with those of all its causes, it is easy to show, as I do, that there can be no causal loops – a fact which on my theory of time is embodied in time's local and global linearity.

Still, of course, as one man's *modus ponens* is another woman's *modus tollens*, you can always infer from this not that causation – and therefore time – must be linear, but that my four chances need not be Humeanly independent. But why do that, when no one believes in causal loops anyway? So why not accept the best possible explanation of their agreed non-existence: namely, a credible and independently motivated theory which makes them impossible – thereby, as a bonus, explaining why time must have the linear topology we all think it does have – a topology which is after all by no means obviously contingent.

JOHN PERRY

Finally, I turn to John Perry, to whose work my theory of time owes a lot. I first met John in Stanford in the late seventies, when I was working on *Real Time I* and he was writing his seminal "Problem of the Essential Indexical" (1979). That was the work which showed me why we agents need irreducible A-beliefs, which any acceptable B-theory of time must therefore let us have.

(My only regret about our discussions then, by the way, was my failure to persuade John to replace his supermarket trolley example with a *Winnie-the-Pooh* story of Pooh and Piglet finding that the snow tracks they've been nervously following round a spinney were in fact made by themselves. He declined, on the grounds that the story wasn't well enough known to American philosophers; but it would have been, if he'd used it.)

I was therefore not at all surprised to find myself agreeing completely with John's refutation of Frank Jackson's (1986) knowledge argument for non-physical facts, and of its temporal and spatial counterparts. About those comments of his I have nothing to say – except that he's dead right. So I hope he will forgive me if I don't yet accept his positive account of knowing what experiences are like. For I still prefer the view of Laurence Nemirow (1990), David Lewis (1988) and others that knowing what experiences are like is not knowing *that* something but knowing *how* something, namely how to imagine, recognise and recollect such experiences.

One of my reasons for this preference, which I've given elsewhere, is our inability to say in non-demonstrative terms *what* we know when we know what experiences of a certain kind are like. In saying this I don't of course mean to deny that we

> have words for properties of experiences, like a "loud" noise, a "sweet" taste, a "warm" feeling: we can say a lot about our experiences. Yet we can say nothing about what they are *like*. What does middle C sound like, sugar taste like, warmth feel like? We cannot say. All we can say is that these experiences are more or less like, i.e. similar to, certain other experiences. But that does not tell us what, in the relevant *non*-relational sense of "like," any one of a set of similar experiences *is* like. This inability to say what our experiences are like would be a complete mystery if knowing what they are like were knowing ... that certain propositions are true (Mellor (1993), p. 8).

Whereas if this knowledge is a kind of know-how our inability to say what we know is not mysterious at all. "In this respect it is like knowing how to ride a bicycle. I cannot state the [proposition] I know then either, because there is no such [proposition] to state." And similarly, I believe, with my knowing what it's like to see red. I think the reason I can't state the propositional content of that knowledge is simple: it has none. This is why I think that even non-physicalists like me should accept a know-how theory of knowing what experiences are like. While to an admittedly "antecedent" – i.e. religiously dogmatic! – physicalist like John, since a know-how theory also postulates no non-physical facts, it should at least provide an acceptable backup to his preferred physicalist know-that theory.

These two theories of what experiences are like also illustrate a general point that I owe to my Cambridge colleague Peter Kail, and with which I'd like to finish, since it rarely gets the attention it deserves. The point is that so-called "anti-realist" or "projectivist" theories of various subject matters come in at least two, quite different, kinds. Some, like the know-how theory of knowing what it's like, deny that our apparent knowledge of an apparent subject matter consists of beliefs at all, never mind true ones. Others, like John's theory of knowing what it's like, or my B-theory of our A-knowledge, do take that knowledge to consist of true beliefs: they just deny that those beliefs are made true by corresponding facts, whether those be non-physical facts about qualia or A-facts about things being past, present or future.

These theories are therefore anti-realist only in the sense of denying that the apparent subject matter of these apparent beliefs exists. They are not anti-realist in the sense of denying that these apparent beliefs are *beliefs*, with distinctive contents, and with the truth values that everyone thinks they have. So these theories in general, and my B-theory of our A-knowledge in particular, are neither non-cognitivist nor error theories. We do not err in believing now that this meeting is present, was future and will be past: those token A-beliefs of ours are all true. The only error lies in the view of Mark and his fellow A-theorists that these beliefs are *made* true by the A-facts that this meeting has the properties respectively of being present, of being future-in-the-past and of being past-in-the-future. This is why I agree with John that we can deny the existence of all such A-facts without denying any of the A-knowledge that we all obviously have.

NOTES

[1] Whether Fred's shot causes or only affects Vicky's death depends in my theory on whether her death would be the same event if it coccurred later, i.e. on whether when this (or any other) event occurs is an essential fact about it (see chapter 12 of *The Facts of Causation*). I doubt it, but need not argue the point here, since the difference – if any – between essential and inessential facts about particulars makes no difference to their *causation*.

[2] And for Davidson's events, for which I see no need to say in general how far their counterfactual identity depends on when or where they occur: see footnote 1 above.

REFERENCES

Davidson, D. (1967), "Causal Relations," *Journal of Philosophy* 64: 691–703.

Davidson, D. (1969), "The Individuation of Events," in N. Resher (ed.), *Essays in Honor of Carl G. Hempel* (Dordrecht: D. Reidel): 216–34.

Jackson, F. (1986), "What Mary Didn't Know," *Journal of Philosophy* 83: 291–295.

Lewis, D. (1990), "What Experience Teaches," in W.G. Lyncan (ed.), *Mind and Cognition* (Oxford: Blackwell): 499–519.

Mackie, J. L. (1974), *The Cement of the Universe* (Oxford: Oxford University Press).

Mellor, D. H. (1993), "Nothing Like Experience," *Proceedings of the Aristotelian Society* 93: 1–16.

Mellor, D. H. (1995), *The Facts of Causation* (London: Routledge).

Mellor, D. H. (1998), *Real Time II* (London: Routledge).

Menzies, P. (1989), "Probabilistic Causation and Causal Processes: a Critique of Lewis," *Philosophy of Science* 56: 642–663.

Nemirow, L. (1990), "Physicalism and the Cognitive Role of Acquaintance," in W.G. Lycon (ed.), *Mind and Cognition* (Oxford: Blackwell): 490–499.

Perry, J. (1979), "The Problem of the Essential Indexical," *Nous* 13: 3–21.

Part Two

TEMPORAL BECOMING

9
PARMENIDES' COMPLETE REJECTION OF TIME

Ronald C. Hoy
California University of Pennsylvania

Parmenides is often credited with discovering the category of timeless truths, and he is sometimes praised or blamed (along with Plato) for asserting that what is real can transcend time.[1] But besides positing a timeless reality for eternal truths to be about, Parmenides finds fault with beliefs about time and argues that time is not real: if temporal thoughts are inherently contradictory then reality cannot be temporal. In claiming time to Plato, Kant, and J.M. McTaggart) who find something unreal about time.

Today, it is easy to ignore Parmenides' rejection of time. There may be timeless truths, and those of logic and mathematics are still enchanting, but they are rarely so compelling to tempt people to deny the irrepressible flux of their experience, especially perceptual experience. It is obvious time is real. So, if Parmenides forces a choice between atemporal, rational posits and temporal experience, many will conclude so much the worse for analytic metaphysics. In the spirit of Henri Bergson and Martin Heidegger, time is real and, if knowable, known in some intuitive way. For others, the response need not be so contentious: Is not today's logic sufficiently sophisticated to provide simple diagnoses of what must have been Parmenides' faulty reasoning? Did he not just fail to appreciate the logic of tenses? Or was he not merely trafficking in semantic misdirection when he worried so much about "nothing," that is, when he denied all attempts to think "what is not is?"

In this paper, I shall suggest it is wrong to interpret Parmenides' position as hinging mainly on semantic issues centered on reference. I shall show how commentators who do so fail to do justice to his complaints about time. Instead, I shall reconstruct Parmenides' worries in terms of the recent conflict between "tensed" and "tenseless" views of time. From this perspective, Parmenides offers an early proscription on the contradictory beliefs that dog any metaphysics based on temporal becoming. It will also become clear how complete Parmenides' rejection of time was: why, for

L. N. Oaklander (ed.), The Importance of Time, 105–129.

Originally published in *The Journal of Philosophy*, XCI, 11 (1994), 573–598. Reproduced here by permission of The Journal of Philosophy, Inc..

other reasons, he could not accept even the tenseless view, and why he should be suspicious of attempts to read him as discovering a new kind of "eternity."

I

Parmenides is often introduced as an uncompromising rationalist, someone willing to abandon common sense to champion an a priori insight. According to F.M. Cornford, (1939), whereas "Heraclitus is the prophet of a Logos which could only be expressed in seeming contradictions, Parmenides is the prophet of a logic which will tolerate no semblance of contradiction" (1939, p. 29). People tend to believe what they perceive, however, and what they believe they perceive are things changing in time. Parmenides' final judgment is harsh: mortals are "dazed, undiscriminating hordes." Carried along by experience, they believe "that to be and not to be are the same and not the same." Wandering "two-headed," the path taken by them is "backward-turning."[2] Although Parmenides does not mention him by name, it is common to read such remarks as warning against Heraclitus' identification of opposites and against his exaltation, in contradictory epigrams, of the backward-turning cycles of fire and war. And it is common to see Parmenides as empowering pure, logical thinking as able to overrule the testimony of perception. Beliefs shaped by perception are alleged to be faulty because they are contradictory, and what is wrong with contradictions, to use Parmenides' path metaphor, is that they are "backward turning." If one asserts and denies the same thing, one does not get anywhere. Contradictions try to affirm and deny, so they really do neither. It is Parmenides' strategy to root out basic contradictions in common beliefs, thereby revealing the true character of reality. If, for example, he can expose as contradictory all beliefs that treat what is real as changing or as temporal, then he can conclude reality is neither changing nor temporal. So far, this sketch of Parmenides may be clear and relatively uncontroversial. What is not clear, and should be controversial, is exactly why Parmenides thinks time is contradictory.

Parmenides[3] (in the voice of a guiding goddess) asserts there are, ostensibly, two basic approaches to reality:

> The one, that it is and that it is impossible for it not to be, is the path of Persuasion (for she attends on Truth). The other, that it is not, and that it must necessarily not be, that I declare to you is a wholly indiscernible track; for thou couldst not know what is not that is impossible-nor declare it, for it is the same thing that can be thought and can be. (Guthrie (1965), pp. 13-14)

Scholars still debate the meaning of this passage, but, at a simple level, Parmenides indicates two apparent paths and says the first one must be taken because the second one turns out to be "indiscernible" and "impossible." But then the goddess indicates what might seem to be a third path, one that involves a failure to appreciate the difference between the first two. It is this latter dead end that is the confused, wandering way of the dazed and undiscriminating:

> [I also hold you back from the way] on which mortals wander knowing nothing, two-headed; for helplessness guides the wandering thought in their breasts, and they are carried along, deaf and blind at once, dazed, undiscriminating hordes, who believe that to be and not to be are the same and not the same; and the path taken by them all is backward-turning. (KRS (1983), p. 247)

It is clear Parmenides believes only the first path should be followed – it is the only path that really leads anywhere. It is not so clear, however, what it is, exactly, that makes the alternatives impossible. That is, there is disagreement about what is responsible for their alleged contradictory character. Some confusion stems initially from the ambiguity of "it" in the indication of the paths "it is" and "it is not."

At the start, it is safe to assume "it" refers to whatever it is that is real. By the end of Parmenides' brief work, he has explicitly affirmed that reality is an undifferentiated unity. So it is reasonable to conclude, by the end of the journey, that "it" refers to such a single entity (sometimes named *being* or *the One).* It is unwise, however, to assume "it" functions as a pronoun for some superentity, being, from the very beginning. Such an assumption tempts one to try to read Parmenides backward: if "it" names being, the phrase "what is not" (in fragment 2) must purport to refer to something else – call it *nothing.* Then, since the path "it is not" is rejected because it is an example of trying to think "what is not," Parmenides must be rejecting the path "it is not" because he thinks it is the same as attempting to think "nothing is." It is at this point that commentators are too often content to focus their analytical and diagnostic skills on the issue: What is problematic about "nothing is"?

Now, it may be true Parmenides would agree that nothing and the void (where these are taken to be pseudo entities ostensibly named by "nothing" and "void") could be examples of what some mortals might believe "what is not" to be, but it should not be assumed Parmenides' talk of "what is not" is, right from the start, an attempt to proscribe such problematic reference. It should not be assumed that Parmenides is, right from the start, equating "what is not" with dubious reference to nothing. He wants to argue, for example, that temporal becoming is impossible. In so doing, he presents genesis and ceasing to be as examples of what is not. But it is far from clear that they are objectionable (or that Parmenides thinks they are objectionable) in exactly the same way ostensible thoughts of nothing might be. Unfortunately, commentators who assume, right from the start, that "it" names being *and* "what is not" names nothing (or nonbeing) are too often content with facile interpretations of Parmenides' denial of "it is not," interpretations that are of little or no help in understanding his denial of the reality of time.

Here is a quick sample of stereotypical diagnoses. According to F. Copleston (1962), Parmenides' point is that "nothing cannot be the object of speech or thought, for to speak about nothing is not to speak, and to think about nothing is the same as not thinking at all" (1962, pp. 66ff). On Bertrand Russell's (1945) interpretation, Parmenides' prohibition Lion against thinking "it is not" is linked to the denial of change as follows:

> When you think, you think of something; when you use a name, it must be the name of something. Therefore both thought and language require objects outside themselves ... [so] whatever can be thought of or spoken of must exist at all times. (1945, p. 49)

Compare Russell's remarks to some of Heidegger's (1961),[4] ones that might serve as a caricature of traditional logic-based philosophy's conception of its roots in Parmenides:

> He who speaks of nothing does not know what he is doing. In speaking of nothing he makes it into something. . . He contradicts himself. But discourse that contradicts itself offends against the fundamental rule of discourse (logos), against "logic." (1961, p. 19)

Such quick equating of "it is not" with attempts to assert "nothing is" run the risk of trivializing Parmenides' position by making it appear as if his reasoning rests on some kind of semantic gimmick. Thus Copleston complains that, when Parmenides denies one can think "it is not" (that is, in Copleston's reading, think of nothing), he is just committing a simple fallacy of equivocation: he is not recognizing (or, he is assuming one will not recognize) the difference between thinking of nothing in the sense of having a thought that negates the existence of things and thinking of nothing in the sense of not exercising one's cognitive faculties at all. Clearly, one cannot think of nothing in both senses at the same time. There is no reason, however, for thinking this trivial confusion is the basis of Parmenides' warning against trying to think what is not. More to the purpose here, there is no reason for thinking such equivocation contributes to what is objectionable in thinking reality to be temporal. Manifestly, in thinking of things as past or future or as becoming, one is exercising one's cognitive faculties and ostensibly thinking of *something*. What is needed is an interpretation that shows why having temporal beliefs commits one to thinking "what is not" in some plausibly controversial way.

Russell does not do much better. According to him, Parmenides is initially agitated by a semantic puzzle: whatever is thought of must be named, so one cannot think of nothing without naming it, thereby turning it into something and failing to do what one set out to do. This suggests Parmenides is just trying to help one avoid the kind of mistake made by the Red King in Lewis Carroll's (1963) story: if nobody passed the messenger on the road, then somebody (namely, nobody) should have arrived first. Supposedly, then, you are urged not to try to think of nothing (what is not) because to do so you would have to turn it into something. But if this is all there is to Parmenides' worry, then it looks as if he gives a semantic trick some legitimacy. Either he is made to look silly (like the Red King) for not seeing it is just a trick, or he is made to look superficial for trying to base rationalist metaphysics on a gimmick. Such semantic interpretations do not clarify, nor do justice to, Parmenides' objections to time and change.

Thus, Russell goes on to read Parmenides as believing that whatever can be thought about (or named) must exist at all times, because it might be thought about at any time. If everything named obtained all the time, temporality would be undermined because all "times" would be the same time (by the identity of indiscernibles). With no change, there would be no time. This line of thought is congenial to advocates of relational theories of time (like Russell), and early philosophers did link time and change. But there is no reason to suppose Parmenides believed reference is to what exists at all times just because *he* supposed successful reference requires the present existence of the referent. Nor does his rejection of time follow from his denial of change (plus the identity of indiscernibles). Rather, it will be suggested below, his denial of change hinges, in part, on his rejection of time: change is impossible, in part, because temporal becoming is impossible.[5]

If Parmenides' rejection of time involves forswearing all temporal concepts (and it does), he would reject the premise that what is thought about (or named) must exist at *all* (present) *times*. And surely he was aware that people typically believe they can talk and think about past or future things without thinking they presently exist. Rather than reject such thinking by implausibly assuming all successful reference is to what presently exists, Parmenides can be read as charging that temporal thinking fails because it is contradictory: time is not real because temporal becoming requires one to affirm that what is ostensibly real *both* is *and* is not. If this case can be made, the kind of semantic thesis Russell suggests as a premise may be closer to being a Parmenidean conclusion.[6]

II

Suppose "it" in Parmenides' "it is" initially refers to whatever exists or obtains, whatever it turns out to be or to be like. (Like a detective arriving at the scene of a crime might say, "whoever did this is a real criminal.") Then, Parmenides'starting point can be taken to be the tautology, "whatever exists, exists." As W. K. C. Guthrie points out (1965, pp. 14-16; cf. KRS 245 ££), it is not cogent to complain that Parmenides begins with a tautology if he is trying to show how people unwittingly contradict it. And being contradictory is what marks the wrong path as wrong right from its start: "that it [namely, what is] is not ... is a wholly indiscernible track." Only after marking this contradiction as indiscernible does Parmenides say the same thing that can be thought can be.

Philosophers from Plato to Russell and Heidegger may have been tempted to read this short equation linking thought and being as epigrammatic semantics: Must every real thought have a real (present) referent? But if, in decrying the indiscernibility of "it [what is] is not," Parmenides is emphasizing the self-defeating character of contradictions, then the test for the genuineness of thought is just its noncontradictory character. It is at least a necessary condition for true thought that it be noncontradictory. Today, it is implausible to think it is also a sufficient condition, since it is a commonplace that many otherwise consistent sentences can nevertheless fail of reference or be false. It should be remembered, though, that Parmenides concluded nearly all mortal beliefs involve contradictions, and he found the set of genuine thoughts to be astonishingly small. Although he may have been unclear about the character and ontological weight of genuine thoughts, and although he may have been wrong about the extent of contradictions, this should not detract from his desire to unmask impostors.

Contradictions fail to be genuine thoughts not so much because they are semantic failures but rather because they are syntactic failures. As such, they fit Parmenides' path metaphors better than semantic concern with naming. To attempt a thought like, "whatever exists (or obtains) does not exist (or obtain)," is akin to stepping in some direction and stepping back. Such a "path" fails to take one anywhere. Parmenides can warn against contradictions without committing himself to either a theory of cognition or a theory of meaning. He can do so without articulating a formal principle of noncontradiction. All he need do (and all he does) is exhibit a basic contradiction (like "what is is not"), remarking such a path cannot be taken.

Accordingly, a minimal interpretation of Parmenides' admonition against trying to think what is not will not portray him as committed, from the beginning, to any prejudice against naming the void, for example, nor will it allege any semantic aversion to a term like "nothing." "Nothing exists" is not objectionable because it requires a pseudo referent for "nothing." Rather, utterances like "what is not exists" are nonstarters because they lead one back to obvious contradictions like "what is is not." (For example, if what is not exists, then what is not is what is; so what is is what is not, and, then, what is is not.) Of course, it is impossible to recover Parmenides' original reasoning, and it would be a different project to try to do justice to the extensive scholarship that tries. Here, working with this minimal starting point has the virtue of forcing an interpretation of what, exactly, Parmenides might have found contradictory about time. It will become clear that, even if Parmenides would endorse, as a general recipe, the identity of what can be thought and what can be, he does not *use* either a theory of *nous* or a theory of reference to reject time. Time is not rejected because one cannot cognitively entertain becoming, or, because one must always refer to what exists now. The motivation and argument go the other way: the appearance of becoming and reference to past and future things are not on the right path because mortals' temporal utterances can be exposed as "backward turning" and "two-headed."

III

What makes them backward turning? Parmenides indicates temporal thoughts involve trying to think what is not is, but, again, it is not illuminating simply to treat this as involvement with the semantically problematic "nothing exists." Again, for example, in thinking of past events as "what is not" it is obvious one is not thinking nothing (in the manner of Copleston), neither in the sense of thinking of the void nor in the sense of failing to have any thoughts at all. And, contrary to Russell, one is not in philosophical trouble just because one is failing to refer to something which is present or which always exists. Ostensibly, one is claiming successful reference to what existed, and this is rarely thought to be nothing.[7] So, what exactly is wrong with believing there are, for example, past (or future) things, things that have or will become?

Parmenides rejects time (and change) in a series of brief remarks in fragment 8. Though brief, they can be viewed as systematically erasing the marks of temporality. (1) Neither becoming nor ceasing to be are real; that is, it is impossible for there to be any genesis or annihilation of what is. (2) What exists is motionless and unalterable. (3) Such temporal terms as "becoming," "was," and "will be" are mere "names" that people erroneously use to try to describe reality. The minimal interpretation suggests what is wrong with believing reality to be temporal is that in so doing one must suppose *what is* also has the status *what is not.* People who believe becoming is real must believe there *are* things (past and future things) that *are not,* making it look as if not to be is the *same* as to be. Yet these people *also* want the being of such things *not to be the same* as things that are (cf. KRS (1983), p. 247-248). All this is two-headed, doubly backward turning.

The argument emerges in the claim that there is no birth of what is real. Parmenides denies there can be any original or ongoing genesis of what is:

> It never was nor will be, since it is now, all together, one, continuous. For what birth will you seek for it? How and whence did it grow? I shall not allow you to say nor to think from not being: for it is not to be said nor thought that it is not; and what need would have driven it later rather than earlier, beginning from the nothing, to grow? Thus it must either be completely or not at all. Nor will the force of conviction allow anything besides it to come to be ever from not being. Therefore justice has never loosed her fetters to allow it to come to be or to perish, but holds it fast. And the decision about these things lies in this: it is or it is not. But it has in fact been decided, as is necessary, to leave the one way unthought and nameless (for it is no true way), but that the other is and is genuine. And how could what is be in the future? How could it come to be? For if it came into being, it is not: nor is it if it is ever going to be in the future. Thus coming to be is extinguished and perishing unheard of. (KRS (1983), pp. 249-250)

In this passage, several objections are quickly hurled at coming to be and perishing. How should their importance be ranked? The most familiar one involves the idea that only nothing could come from nothing. Some commentators focus on Parmenides' apparent use of a principle of sufficient reason, as if his primary objection is to a cosmological picture of creation from nothing. W. I. Matson's (1987) interpretation of Parmenides on this point is that "every real thing must have a real cause ... reality cannot just pop up out of nothing – nothingness by definition not being a real thing;" in a Leibnizian vein, "supposing that Being could come from nothingness, there could be no reason why this event should occur at one time rather than any other - so it could not occur" (1987, pp. 31-32). But such causal reasoning should take a back seat to Parmenides' point that trying to think of things coming to be involves the impossible "it [what is] is not." His use of a common causal belief, like nothing comes from nothing, should be put in the context: "even if it were possible for there to be coming to be, only what is not could come to be from what is not." What is wrong with this is that it produces the contradiction "what is not is." What needs to be made clear is why belief in becoming involves the contradiction "what is is not."

The following reconstruction highlights the link between becoming and this basic, forbidden contradiction. It takes its cue from the objection that, if what is came into being, or "is ever going to be in the future," then it "is not." In other words, if what is came into being, then there was a time when it "is not" – namely, when, in the past, it was in the future. And if what is "is ever going to be in the future," then what is "is not" – namely, now. Instead of focusing on the causal impotence of nothing, or on semantic tricks with the term "nothing," this reconstruction elaborates the *contradictory* character of becoming:

(1) To try to think of what is becoming (or the genesis of what is), is to think there is some past (to precede what becomes), and, it is to think there is some future (from whence what becomes issues) – that is, it is to think the past and future are real, part of what is.
(2) But mortals also say the future is what is not (and, what is future is what is not).
(3) So, to try to say there is some future, or to try to think of what is as future, is to try to affirm both what is said to be what is and what is said to be what is not (namely, the future).
(4) It is impossible (and forbidden) to think what is is not (or, what is not is).
(5) Therefore, no future is really thinkable.
(6) Since what is must be thinkable, there can be no future.
(7) Similarly, mortals also say the past is what is not (or, what is past is what is not).
(8) So, again, to try to say there is some past, or to try to think of what is as past, is to try to affirm both what is said to be what is and what is said to be what is not (this time, the past).
(9) Again, this is impossible, and there can be no past.
(10) Therefore, there can be no becoming (or coming to be or genesis) of what is.

This reconstruction presents the way people think about becoming as committing them to treating the past and the future in contradictory ways. For what is real to come into being, it must have been future and there must be a past when what comes to be was future. Insofar as people think there is some past and there is some future, they try to put them (and what is in the past or future) on a par with what is – treating to be and not to be the same. Parmenides objects to this kind of backward turning and more: people *also* claim the past and future are different from what is (after all, the past and the future are *not* what is now). Such thinking is what Parmenides condemned when he said, "[mortals are] dazed undiscriminating hordes, who believe that to be and not to be *are the same and not the same*" (KRS (1983), p. 247, my emphasis).

Primarily semantic readings of the complaint about coming to be, focusing as they do on the void and creation, can plausibly say how nothing is treated as the same as something (or nonbeing the same as being), but how do they account for the other part of the complaint – that to be is believed to be the same *and not the same* as not to be? If one goes so far as to think of nonbeing on a par with being, how are they not the same? Rather than fixate on the void and cosmogony, the above reconstruction of *typical* thinking about ostensibly *ongoing becoming* gives Parmenides' complaint some bite. In becoming, what is is supposed to be different from what is not, but also, and again as a result of becoming, the same as what is not.

Another virtue of this reconstruction, compared now with those emphasizing an a priori principle of causality, is that it can explain why, practically in the same breath, Parmenides attacks both becoming and perishing. Perishing (or ceasing to be) involves becoming and the contradictory character of what is past. Continuing the argument:

(11) For what is to perish (or cease to be), what is must come to be past.
(12) But there is no coming to be (10), and there is no past (9).
(13) Therefore, what is cannot perish.
(14) "Thus coming to be is extinguished and perishing unheard of."

Causality (or sufficient reason) readings are typically quiet about why Parmenides denies what is can cease to be. No such principles (as opposed to conservation principles) are violated if what is ceases to be. But focusing on becoming provides an answer: besides being "dazed" and "undiscriminating" when imagining the future as a womb where what is is what is not while it waits to be born, mortals are similarly "two-headed" when they imagine the past as a resting place for what is to be what is not. Again, the problem is not that the past is nothing; it is not that people think of the past as containing nothing and are thereby duped into thinking it contains something. Rather, the past is thought to be *and* not to be. And it is thought to be loaded, but with impossible things – things that are *and* are not.

Later readers have generally failed to appreciate the extent to which Parmenides' attack is, from the beginning, an assault on temporality rather than a mere rejection of genesis as an initial cosmic event. Influenced by religious stories of creation, or by scientific theories of a "Big Bang" preceded by vacuum field states, readers may easily be tempted to see Parmenides as mainly concerned to deny the genesis of the universe from a prior void or nothingness. It may be true that Parmenides rejects the void, but his rejection of "coming to be" is not just a denial of an initial creation event. Nor does it rest on his denial of the void. Actually, it was typical for ancient Greeks to think of the void as what *has* the potential to become, and, if so, Parmenides could be rejecting the void *because* becoming is impossible –

what is must "fully be." Thus, contrary to the usual tendency to see Parmenides' rejection of time as based on some semantically inspired aversion to the void (or nonbeing or whatever), his rejection of the void may have been motivated by what he took to be the contradictory character of coming to be. And his rejection of coming to be is more general than a denial that what is could come from nothing. There cannot be *any* becoming (or perishing) because to suppose that *whatever* is real (even a hypothetical void) undergoes becoming (or perishing) requires the impossible conflation of what is with what is not.

IV

Logic has become more sophisticated since Parmenides' time, and so it will appear to many that becoming, perishing, and time can be rescued from this attack. In particular, paying careful attention to the logical distinctions signaled by grammatical tenses are supposed to allow one to avoid saying what is both is and is not. Indeed, with the assistance of logical quantifiers one should even be able to envision the possibility that what is came to be subsequent to a state of nothingness. Would Parmenides be persuaded?

Consider first the benefits of grammatical tenses. Tenses are a way to indicate whether what is talked about is past, present, or future. You *read* the sentence before this one – the one you *are now reading* and this, in turn, is happening before you will *read* the next sentence. Here, the differences between "read," "are reading," and "will read" help make it clear you read the different sentences at different times. This forestalls the contradiction that you both read and do not read a given sentence by making it clear these different states of affairs obtain at different times. When Parmenides says things like "what is, is" and "it is impossible to think what is is not," it is natural to suggest that "is" should be read as a tensed verb. Then, it is tempting to accuse Parmenides of failing to appreciate the use of tenses to describe time in a noncontradictory way.

Thus, what is now exists, but where is the contradiction in saying what is now *did* not exist? Once upon a time, what is now was future, and when it *was* future then it was what is not. When it was true that what exists now *was* what is not, then it would have been wrong to refer to what exists now as "what is." Instead, it should then have been referred to as "what will be." Similarly, at some future time, what is now will be what does not exist. What is (now) will be what is not (then). What is now will be what *was*. Then, what is will be past. Then, it will be wrong to refer to what is now as "what is." Instead, it should then be referred to as "what was." In this way, one might hope sufficient marks have been invented so the coming to be and perishing of what is can be tracked without losing one's way on the "what is both is and is not" path.

Also, logic can show how to conceive the possibility that what was once true was the proposition "what is is nothing." Parmenides is supposed to have rejected such a thought because he took it to be equivalent to the contradiction "what is is what is not," or because he worried one cannot try to think of nothing without transforming it into something, the referent of "nothing." By means of its analysis of general (quantified) sentences, modern logic suggests such contradictions and semantic puzzles might be avoided by finding a coherent paraphrase. In particular, suppose "what is is what is not" is intended to mean the same as "nothing exists." Instead of paradoxically reading the latter as talking about something ("Nothing") existing, logic offers the paraphrase: "For all things, x, it is not the case that x exists." Here, it is stipulated "x" holds a place only for genuine names. Now, if "nothing" were *erroneously* treated as a name, then substitution of "nothing" for "x" would result in the opposite of what was originally intended: "it is not the case that nothing exists." But logic insists "nothing" is not a name; instead, it is a term signaling what is here meant is that *no* names will make the open sentence "x *exists*" true. Thus, so far as logic is concerned, there is no contradiction or semantic puzzle if one says, "What is true is that nothing existed." All this need mean is, for all would-be names, none of them would make the open sentence *"x existed"* true.

Combining the resource of tenses with the logic of general existence statements leads to an interpretation of the following table as a picture of becoming. It is a scheme that includes not only becoming, but also genesis from nothing and the perishing of what is.

Table 1

	t_0	t_1	t_2
what is	[void]	t_1-things	t_2-things
what is not	future things	past void & future things	past void, t_1-things *& future* things

The interpretation of this schema can be embodied in a further table showing the kinds of things that would be true at different times:

Table 2

t_0	t_1	t_2
now true	*now true*	*now true*
nothing exists	nothing existed	nothing existed (before t_1)
t_1-things will exist	t_1-things exist	t_1 things existed
t_2-things will exist	t_2- things will exist	t_2-things exist

Criticism of Parmenides' denial of the reality of becoming, of genesis from nothing, and of perishing can be summarized in terms of these tables. If his objection to "it is not" is based on contradiction causing substitutions of "what is" for "it," then tenses provide a way to avoid this substitution: substitute instead "what was" (or "what existed") or "what will be" (or "what will exist"). Table 2 shows how one can try to insist that what is is just what exists in the now, which is signaled by the appropriateness of the present tense. And the table shows how one can talk about what is not (if it is not because it is not now) by shifting to the correct future or past tense. Additionally, if Parmenides believes what is is a single, undifferentiated entity – a kind of superobject and if his objection to genesis is grounded on the idea that to think about the genesis of this superobject requires positing a pseudo object (nothing) as prior to the superobject, then one can see how to avoid reference to pseudo objects. All one need say is what was, once upon a time, was the complete absence of what will exist at any later time. In other words, no nameable thing existed. At later times, one can say, "it *was* the case that, for all x, x does not exist" (remembering that "nothing" and "void" are not allowed to be names).[8]

V

How might Parmenides respond to the tense defense of becoming and perishing? Since he cannot be asked, listen instead to "Tenseless Martina," an apprentice of his featured goddess. Tenseless Martina speaks:

"We enjoy sophistication in logic and grammar, but we must be sure it takes us along the right path. "Advances" may dazzle some mortals into thinking they have so progressed, when in fact they are just getting more dazed and two-headed. Alas, the dizzying dance of tenses only postpones the realization that mortals treat what is not as if it is and what is as if it is not.

"First, however, let me comment on the explanation of the possibility that, once upon a time, nothing exists. Offered as substitute for "nothing exists" is the alternative, "for all x, if x is a thing (or object), then x does not exist," where it is stipulated only names of would-be things can be substituted for x. This is a marvelous offering which in no way conflicts with our understanding of the firm truth "whatever is, is." Mortals believe in the existence of lots of things. They use many names that they are willing to substitute in place of x; this shows they believe "nothing exists" is false. We, however, have reasons for denying that such "things" – the objects of the phenomenal world of mortals – are real. This may surprise you, but we are actually willing to affirm the truth of your sentence "nothing exists," if all you allow to be substituted for x are names of the things ordinary mortals think are real. "What is, is" is compatible with none of these things being what is. It is not we goddesses who are worried that, if no such temporal objects exist, then some paradoxical void, nothing, or some such is what is. It is mortals who say strange things about nothing, and it is mortals (like Plato and Russell) who show an irresistible urge to demand referents for nearly all their sounds.

"We prefer to say simply: whatever is, is. *We* are not upset if what is is not an object. Ours is the way of truth – not the way of superobject. *We* are comfortable with the correlation: what is (or obtains or is the case) is what makes truth. We still insist, though, that one avoid the path whereby what is not *is* (or what is not true is true). If one of my sisters, in her enthusiasm, seems to name what is (saying, "being") no great harm is done. In this case, you may conclude "for all x, x does not exist" is false, but just for the trivial reason that "being" names whatever is, and so substitution in this case would yield something equivalent to the contradictory "whatever obtains does not obtain." Thus, if "being" names whatever obtains (even the state void of mortal things), then it must be true that "nothing exists" is false – according to the analysis offered. (Notice, too, we are not playing games with the word "nothing;" we are not saying "nothing exists" must be false because "nothing" must name something to be meaningful.) *We* are happy with this result, but mortals may not be. Because, in spite of these extensions of logic, it may be that what "being" names is quite different from any of the things for which mortals have names. Trying, then, to imagine such a case and failing, mortals might also jump to the contrary conclusion that, after all, "nothing exists" is true, too. They would be wrong, but not for the silly reason that they (or we) are attempting to use "nothing" as a name. They would be wrong just because they would not be letting "whatever is, is" be true. They may not have forgotten being, but they will have forgotten they let my sister use "being" as a name for whatever is.

"So, I allow you to make the sounds "nothing exists" syntactically safe for predicate logic and for a science of objects named by mortals – if this is what you want to do. And by all means, continue to deny the word "nothing" the status of a name! But do not think that by giving coherent sense to "nothing exists" you thereby make it possible to negate "whatever is, is."

"Now, on to the incredible powers of tenses. They are like candles in a draft: what they show is aetherial – "all is full of light and obscure night at once" (cf. fragment 9) . In the dark, flickering flames give mortals dancing images of what seems to be. Similarly, tenses portray what is as if it were just a brief passage in darkness. But the passage is a maze, where things said not to be are yet sufficiently located to come to be, and, having been, pass on to a different place where, again, they are and are not. Tenses may help mortals keep track of where, in their wandering story, they think they are. Tenses keep the tale in hand, so to speak, so it seems one's passage goes from a beginning to an end. For the telling, tenses add drama, but we are concerned with the whole story – the whole of what must be said to be.

"The sparkling consistency of tensed logics, in the narrative sphere, cannot hide for long the fact that tense is a path on which "to be and not to be are the same and not the same." To see this, ask (using a twentieth-century idiom), what are the truth conditions for tensed statements? For example, look at what is true at t_1 in Table 2. Suppose you are at that point in the maze. It is easy, then, to say what the truth conditions are for "t_1-things are what is," because t_1 things exist then; the light of the present tense allows *them* to be displayed, to appear. But what about the past tensed "nothing existed," or the future tensed "t_2 things will exist?" The champion of tense tells the story that what is is *only what exists now,* so when such a one is at t_1 he can do little more than point to Table 1 as a collection of what is and what-is-not marks. *But these are just marks, and at t_1 the marks available then are themselves all just t_1-things.* The tense champion cannot use real t_2 – things, for example, to show their location and to make his future-tense sentences true, since, for him, they are not. In desperation, some mortals attempt to conjure hybrid monsters: present "facts" about the future (or about the past). But how could there be present facts about either the future or the past without there *being* future or past things? Even more desperate, some deny there are any truths about the future. But why, then, should there be any truths about the past either? Clearly, this is a path that goes nowhere and has been nowhere. Do such problems not just show that tense is terminally infected with the "two headedness" of the past and the future?

"Mortals of the twentieth century have a more clever way to try to show the reality of time, using what they call a "tenseless analysis" of the truth conditions of temporal beliefs. I have a weakness for this theory, which is why my sisters call me, "Tenseless Martina." Champions of tenseless time have an easier time with the above questions, for they make full use of Table 1 by embracing *all* of its what – is row. In so doing, they affirm that t_0-void, t_1-things, and t_2-things all exist (tenselessly) at t_0, t_1, and t_2, respectively. What makes them different *times* is a long part of the tenseless story, but for our purposes here what is important is simply the idea that, in the tenseless story, things do not come to be or perish in any metaphysical sense – they are. Things exist tenselessly at their temporal locations (not at all temporal locations), and *being* there they can serve as truth conditions for sentences, tensed as well as tenseless, no matter when these occur. For example, a t_2 sentence like the tensed "t_1-things existed" is made true by the existence of t_1-things earlier than the tenseless existence of the t_2-sentence token,

"t_1-things existed."[9] Now, if a tense mortal could relax and see, for a moment at least, that such existence is what makes even tensed sentences true, then he should more readily confess that, after all, *he is* trying to say what is not (namely, what he says is past or future) *both* is what is not and what is. In Table 1, *he* puts the void and t_2-things in the what-is-not row under t_1 – this is where he tries to say how they are *not the same* as what is. But to explain the t_1-tensed truths of Table 2, he also must refer to the void and t_2-things as *being* what is represented by the what – is row of Table 1. To try to hide this two headedness he talks of them as being "in the past" or "in the future" (as if he could add a hybrid row to Table 1), but if he has them *be* anywhere he really tries to make them the same as what is. Remember, he does this while also making them different from what is by putting them in the *what-is-not* row of Table 1.

"Surely, then, the mortal who believes in the reality of becoming is in a maze that inevitably turns back on itself. If such a mortal does not see even so far as tenseless time – by refusing to grant the extension of the what is row beyond the moment of his utterance – how can he conjure the truths needed to sustain becoming? Unless *there is* some future and some past, there is no becoming. Time vanishes if there is only what is at t_1, for example. Some mortals, dazed and desperate, claim to have extraordinary vision: they claim to be able to look into the what-is-not and discern the shadowy being of a past and a future. But this is a trick with mirrors or mushrooms. If the status of what is past and what is future is reflected only in the what-is-not row, then they are not – period. Then there is no distinction between "things" past and future; there is no thing that is coming to be, and there is no thing that is perished. Such things – the things becoming and ceased - are only in mortals' dreams. At this point, the teller of tensed stories might turn again in his maze and grope: "Well, things do exist in dreams, so why can things not exist in the past or future in an analogous way?" But how? Philosophers have made admirable progress in explaining how the "what-is-nots" of dreams, literary fiction, and modal discourse can be understood in terms of just what is. For future and past entities, however, for things yet to come to be or things perished (and not just dreams of them), it is totally obscure how their being what the believer in becoming needs them to be can be reduced to what exists at some other time, some other present. So, can becoming be more than a dream?

"Again, the mortal who believes tenses mark what is from what is not is exposed to be ever two-headed and backward turning. The best a mere mortal can do is admit to equivocation: there exists (tenselessly) what does not exist (tensed) now. To admit this is to admit coming to be and perishing are mere appearance. Tenses do not separate what is from what is not. Instead, they mark off what is from what is – what is experienced or thought (what appears) at one point from what is experienced or thought at another.

"But most mortals are incorrigible. Let them take their tensed tales in hand, run their mazes, and be of the opinion what is becomes and ceases to be. Those of you who would like to try some other path can begin by erasing the what-is-not row of Table 1. Then you are left solely with what is: "coming to be is extinguished and perishing unheard of." Leave tenses in their place – Table 2 – as mere tracks of mortal wandering."

So says Tenseless Martina. For her, tenses only temporarily hide the contradictory character of becoming. But perhaps Tenseless Martina has not been fully forthcoming, for she gives the impression she endorses the reality of tenseless time. It remains to be seen why Parmenides would also reject the reality of tenseless time.

VI

If Tenseless Martina can use the tenseless theory of time to explain how becoming and perishing are two-headed and backward turning, it might seem Parmenides should be able to endorse the tenseless view of time, which says time is real though metaphysical becoming is not. Such tenseless time is in accord with the vision of some physicists and philosophers according to whom the universe is a four-dimensional whole, with earlier-than and later-than relations ordering the tenseless existence of things along one of these dimensions.[10] The appearance of becoming is just a perspectival matter, reflexively involving local, occurrent perceptual or belief states.[11] Different states of being, so to speak, are represented in different states of human beings at different locations in their world lines.[12] If Parmenides' point was that truth must be timeless, perhaps he could accept the view that timeless truths are just tenseless truths – truths that are true at every time. If, then, reality is pictured as a four-dimensional whole with no two-headed metaphysical becoming, could Parmenides accept the reality of time – albeit tenseless time – after all?

There are reasons to doubt Parmenides would accept the reality of any kind of time, including the tenseless time of scientifically inspired philosophies. First, Parmenides rejects the reality of change altogether, while tenseless theories of time analyze change as things being (tenselessly) different at different earlier-than/later-than "locations." Second, tenseless theories of time need reality to be diverse, so differences amongst things and events can ground temporal relations.[13] Parmenides, however, totally rejects diversity.

In early Greek thought, it was common to view change as a species of becoming and to fail to make a clear distinction between substances (individual things) and qualities.[14] An apple's changing from green to red might be described, in Heraclitean fashion: the green apple is and is not; the red apple is not and is. The green apple ceases to be (and becomes what is not) and the red apple (which is not) comes to be what is. Against this background, if becoming and ceasing to be are cited as unreal, it is natural to expect that motion and alteration are automatically in trouble. Even for enduring substances, if one jogs from A to B, his being at B must, it might be thought, come to be after his being at A ceases to be. And the ripening of an apple seems to require the ceasing of its being green and the coming to be of its being red. To the extent any change involves becoming and perishing, Parmenides might reject them all *just* for being cases of becoming or perishing. It is in one breath that his goddess lumps motion and change of color with becoming and perishing:

> ... there neither is nor will be anything else besides what is, since Fate fettered it to be whole and changeless. Therefore it has been named all the names which mortals have laid down believing them to be true – coming to be and perishing, being and not being, changing place and altering bright colour. (KRS (1983), p. 252)

What is important for tenseless time, however, is not change via becoming but change analyzed in terms of diversity. So Parmenides' thoroughgoing monism is the more serious obstacle to his accepting tenseless time. He claims reality is single and undifferentiated. If so, then it cannot be one way (or thing) at one point and some different way (or thing) at another. What is must be "equally balanced:"

> ... it needs must not be somewhat more or somewhat less here or there. For neither is it non-existent, which would stop it from reaching its like, nor is it existent in such a way that there would be more being here, less there, since it is all inviolate: for being equal to itself on every side, it lies uniformly within its limits. (KRS (1983), pp. 252-253)

Although they agree with him about the phenomenal, mortal-relative character of becoming, tenseless time needs variety amongst things and events to be the basis of temporal relations. Parmenides' severe monism, if adhered to, would force the rejection of tenseless time.

VII

Problems also face those attempts to qualify Parmenides' rejection of time which read him as endorsing some kind of "eternal present." Some commentators have strained to construct some kind of eternity as the mode of being for Parmenides' One, even though they have recognized his total denial of becoming and change. Perhaps they have stumbled on his remark that what is "is now, all together." Here, he seems to emphasize the present, indeed the present tensed, character of reality, but it is a mistake to take "now" literally or to make it eternal.

Kirk, Raven, and Schofield translate the provocative remark: "It never was nor will be, since it is now, all together, one continuous." And they comment: "Probably what Parmenides means to ascribe to what it is is existence in an eternal present not subject to temporal distinctions of any sort. It is very unclear how he hoped to ground this conclusion in the arguments of [fragment 8]" (1983, pp. 249-250). The reconstruction in section III above hopefully makes the grounding clearer, but what do Kirk, Raven, and Schofield mean by "eternal present?" How can a present, or what is present, be eternal without its being both *present* and present at all *times*? Neither is possible for Parmenides.

If Parmenides denies the applicability of concepts of the past and future to reality, what concept of *the present* can be left? It was typical, remember, to link time and change, with changes seen as instances of becoming and ceasing to be. The supposition that time might pass though everything remains the same is uncommon. In modem times it has affinity mainly with the Newtonian worldview, where time is mathematically postulated as a constant, absolute flux, independent of the existence of particular objects. Parmenides' view is not Newtonian, so when he denies the reality of change and becoming – when he denies the reality of time – he should hold all temporal concepts to be specious. When becoming is "extinguished," when past and future being are exposed as contradictory, what is left is not "the present." What is left is neither a moment of dynamic becoming nor something temporally distinguished from the past or future. What is left is not now.[15]

Next, without some concept of a temporal present, notions of the eternal become problematic. And on any conception of the present, Kirk, Raven, and Schofield's "eternal present" is in trouble. What is present is supposed to be ontologically ephemeral according to tensed theories of time, and this (again according to tensed theories) is what makes the present a special moment of time. According to tenseless theories, on the other hand, what is present is just some set of events neither earlier than nor later than some event in question. What makes some of these events special (of pragmatic importance) is that they are simultaneous with a conscious token capable of falling under the general linguistic type: "[the event] is simultaneous with this token."[16] Importantly, however, according to neither view of time can a present be present at any *other* time, let alone at *all* times! So no present can be eternal in the sense of *everlastingly* being the same present. If Kirk, Raven, and Schofield suppose what is is "eternally present" just insofar as what is is the same *throughout time,* then *they* would be assuming the cogency of some notion of temporal passage for a reality claimed not to be temporal – as if Parmenides' reality were still subject to Newtonian time. This would be an injustice. If Parmenides recognized that voiding concepts of becoming and change extinguishes time completely, then he should be credited with an early insight – an insight which kept him from saying things like, "what is will be eternally the same in all future nows." When Parmenides abolishes time, what is left is not an eternal present.[17]

Here is another suggestion. Guthrie recognizes Parmenides' claim that the "past and future have no meaning in or for reality" implies "the time sequence is abolished." Unfortunately, he credits Parmenides' with a further achievement: the "recognition. of *eternal* as a separate category from *everlasting* (1965, pp. 26-29). This is unfortunate because it is not at all clear what kind of category the eternal would be if it is to be distinct from the everlasting. Guthrie refers to Plato's distinction between the status of Forms and the everlastingness of the temporal cosmos, suggesting Parmenides' reality may be "eternal" like Plato's Forms. There is a connection between Parmenides and Plato's attempt to boost his Forms beyond time, but Guthrie has it backward.

Plato may have been inclined by Parmenides to suspect what is fully real cannot be subject to time, but he thinks *many* entities – many Forms – are fully real. Moreover, he takes seriously the problem of trying to understand the relation between Forms and the realm in which temporal appearances come to be and perish. Basically, *Plato's* skepticism about time stems from his worry that when particulars change they undergo transitions from exemplifying one Form to contrary ones, and in so doing there are moments when particulars are neither one nor the other. Plato concludes that becoming, time, and temporal particulars are incompletely intelligible; his reasons for avoiding the troubling gaps in exemplification involve his analysis of predication, change, and his belief that no Form is named by "nonbeing." In the end, though, Plato does not deny that changes happen or that time is real; rather, he denies their full intelligibility – their inability to be fully described in terms of Forms. Forms, however, are not infected by change, and being, so to speak, "above the show" (above the "moving image") they transcend time. In Plato's system, it makes some sense to say that Forms are eternally present, namely, the sense in which they are available *at any time* to be exemplified. But this eternal availability does not capture their temporal transcendence, since Plato also envisages Forms existing independently of (metaphysically prior to) the temporal order of particulars. Platonic Forms are themselves supposed to be atemporal in spite of their being eternally present (available) for particulars.[18]

The Platonic scheme does not help explicate Parmenides. Unlike Plato, Parmenides extinguishes time and change altogether. Rather than try to use the eternity of Forms as a model for Parmenides' reality, one should go in the opposite direction: use the *atemporality* of Parmenides' reality as a model for the *atemporality* of Plato's Forms. It may be mysterious, and it may be wrong, but the atemporality of both is meant to be quite different from everlasting presence.

Guthrie's trying to point to some kind of eternity that would be different from the everlasting may strike a sympathetic chord with scholars interested in theological efforts to find a way for God to be eternal which would transcend the imperfections associated with mere temporal everlastingness. But this effort is also problematic and anachronistic: a basically anthropomorphic deity who thinks, wills, creates, and effects miracles in the temporal world was suddenly required (by later philosophically sophisticated theologians) to conform to a Platonic ideal of being perfect and, therefore, *not* temporal. Simply calling such atemporal existence "eternal" does not illuminate this new usage, nor does it show how it is an advance on the concept of the eternal in the sense of the everlasting.[19]

Supposing, then, Parmenides' reality is best characterized as atemporal, what finally should be made of his remark, it "is now, all together?" He may just be emphasizing the completeness of reality. People are familiar primarily with developing things, things that are incomplete, unfinished, or growing. Such things take time to realize their nature – be fully what they can become. Perhaps the closest Parmenides' goddess could come, using the language of mortals, to indicate that reality is *not at all like a developing thing is* to use the popular belief that *only* what is now exists – as a present existent, what is now has ostensibly achieved a momentary completeness unmixed with the past and the future. (Remember, of course, this belief turns out to be hypocritical since mortals also try to grant existence to what is past and future). Then, to say what is "is now, all together" could have the force: what is is *like* what you mortals (in one of your frames of mind) believe present existence to be, namely, unmixed with pseudo entities like the past and future. *Unlike* the momentary presence of states of developing objects, however, Parmenides' what is is *genuinely complete;* it has its being "all together" and alone.

Mortals, of course, would like to know more about what is complete in this remarkable way, marked neither by the what-is-not of a perished past nor by the what-is-not of a merely potential future. The tenseless theorist, inspired by Tenseless Martina, is primed to say it is what the tenseless existence of the world's "career" is like – the events of this career are spread out, yet hang together, along the fourth dimension of a four-dimensional, spatiotemporal plenum. This whole, with its parts, provides (tenselessly) the truth conditions that make tenseless truths eternally true (and tensed statements occasionally true). But this line of thought is a rapprochement with tenseless *time,* and it would require Parmenides to embrace the heterogeneity of what is. Any attempt to spread out Parmenides' One in the empiricists' four-dimensional hyperspace fails as its homogeneity collapses it back into a singularity. Deprived of diversity – the empiricist's basis for spatiotemporal relations – the all-togetherness of Parmenides One has the well-roundedness of a solitary point. Given Parmenides' severe monism, it is futile to try to describe this singular reality as other than completely atemporal. If, however, what is can be diverse, then mortals can go on with the project of explaining how their temporal experiences are a part of reality. For, were it not for his monism, Parmenides' turn away from time need only have been a turn from the two-headed ontology of tensed time; it need not have been a turn from the way of tenseless time.

NOTES

[1] See, for example, G.E.L. Owen (1974, pp. 271-292).
[2] From fragment 6 as given by G.S. Kirk, J.E. Raven, and M. Schofield (1983), p. 247 (hereafter cited as KRS).
[3] Fragments 2 and 3 as translated by W.K.C. Guthrie (1965).
[4] Heidegger, as part of a larger project, wants to talk about nothing in his own way. In this regard, Parmenides is to be admired not because he uses logic to banish what is not, but rather because he is one of the first to realize "that along the way of being the way of nonbeing must be specially *considered,* that it is therefore a misunderstanding of the question of being to turn one's back on nothing with the assurance that nothing obviously is not" (1961, p. 94). When it comes to the "path of nonbeing," however, Heidegger is sanguine: "to be sure it cannot be traveled." And it suffices to point to his phenomenological orientation parenthetically: "(For that nothing is not an essent does not prevent it from belonging to being in its own way)" (1961, p. 94).
[5] Owen also makes the identity of indiscernibles too important and makes Parmenides look as if he assumes what is does not change: if what is is unchanging, then "nothing can be said of it in the past or future tense" (1974, p. 273). Complaining the argument is incomplete, he supplements it with the identity of indiscernibles. Although Owen sees this as anachronistic and recognizes "the familiar readiness of the Greeks to picture the lapse of time as the parent and regulator and assessor of change," he does not explore the likelihood that Parmenides' rejection of time is logically prior to his denial of change.
[6] Philosophers who are quick to find Red King semantics as the key to Parmenides risk an understandable, and perhaps old, anachronism. There is no doubt that problems of meaning and reference have been fertile in the history of philosophy, and Parmenides probably influenced this history by prompting Plato's "semantic" aversion to "nonbeing." But Plato had other concerns, and his reasons for avoiding positing a Form for nonbeing should not be assumed to be the same as Parmenides' aversion to affirming what is not. True, Parmenides says "it is the same thing that can be thought and can be" (fragment 3) and he remarks "nothing is not" there for being or thought (fragment 6). But his reason for denying thoughts of "what is not" may be both simpler and, in the case of time, more provocative than either making or wanting to avoid the Red King's mistake. And even if Parmenides had semantic

concerns, surely there is no reason to think he was constrained by Russell's epistemic requirement that meaning and reference be anchored in the present data of acquaintance!

[7] Some talk of "creation," or origin, of the whole universe may be an exception, if it is assumed nothing, or a void, "preceded" the universe in time. It will be argued in section V that even this encounter with nothing need not be very problematic for Parmenides.

[8] Another possible criticism is that Parmenides overreacted to Heraclitus' habit of apparently *identifying* what exists at one time with what exists at another. Suppose Parmenides wanted to avoid identifying what is with what is not in the way Heraclitus seemed to identify summer and winter. Unless one is talking about special things (like enduring substances), it is wrong in general to identify things existing at one time with those existing at others. Yet this is how Heraclitus tended to describe change, for example, "as the same thing there exists in us living and dead and the waking and the sleeping and young and old: for these things having changed round are those, and those having changed round are these" (KRS (1983), p. 189). Careful use of tenses can put a stop to such contradictory identities.

[9] Tenseless analyses of time have their roots in relational theories of time, theories that hold earlier-than and later-than relations amongst events as more basic, and more legitimate, than any *sui generis* ontological passage or flux. They have affinity with Leibniz's view of time and have been advanced by Russell, Hans Reichenbach, W .V. Quine, and Adolph Grünbaum. Here, Tenseless Martina's comments are in the spirit of D. H. Mellor's *Real Time* (1981).

[10] A connection between Parmenides and space-time physics was noted by Kurt Gödel while discussing the importance of understanding the implications of the relativity of simultaneity: "it seems that one obtains an unequivocal proof for the view of those philosophers who like Parmenides, Kant, and the modern idealists, deny the objectivity of change and consider change as an illusion or an appearance due to our special mode of perception" (Gödel (1959), p. 557). Now, however, tenseless theorists tend to see the special theory of relativity as undermining the assumption there is a universal "now," rather than as denying the objectivity of change.

[11] See Grünbaum (1967, ch. 1) and (1969). See also Mellor (1981, ch. 3).

[12] Tenseless Martina might put it poetically: at each presenting location (at times that include mortal representings of temporal things), Being has beings "presencing" Being. These ecstatic turns of phrase are Heideggerese, but Martina does not care. Heidegger, though, would be shocked by the suggestion of any sameness of doctrine with the tenseless time of analytic philosophy. Much of Heidegger's work can be viewed as poetic transformations of Edmund Husserl's attempt to describe the complex intentional structure of the specious present. (See Husserl's (1966) lectures, which Heidegger edited.) As a phenomenologist, Heidegger gives priority to describing the prescencing of Being from a human perspective. From his point of view, he is adamant: "Whatever is presently present is not a slice of something shoved in between what is not presently present; it is present insofar as it lets itself belong to the non-present" –Heidegger (1975, p. 44). This is work: "What has-been which, by refusing the present, lets that become present which is no longer present; and the coming toward us of what is to come which, by withholding the present, lets that be present which is not yet present – both made manifest the manner of an extending opening up which gives all presencing into the open" (Heidegger (1972), p. 17). For all the hustle, Martina can smash these lobs: "Only tense mortals worry about their presently present consciousness being a mean slice between nonbeing. We who take the tenseless view see them all as moments which are (tenselessly) part of what is." For other worries about time and consciousness, see my (1978), (1982), and (1985).

[13] A theory of temporal relations becomes part of a theory of object identity, of saying why some clusters of more or less diverse property instantiations should qualify as objects, and, why even more diverse sets should qualify as histories or "space-time careers."

[14] Cf. Guthrie, (1965), p. 43.

[15]Something can be said to be present only as part of a network of temporal distinctions. If concepts are "schematized" temporally, as Kant claimed, then to think about something as present is to be able to think about its causal relations. Without corresponding relata, the idea of *present* existence is empty. See the "Anticipations of Perception" and the "Second Analogy" sections of Kant's *Critique of Pure Reason.*

[16] Cf. Mellor (1981, chs. 2 and 3).

[17] As further evidence that Kirk, Raven, and Schofield miss the completeness of Parmenides' rejection of time, they worry: "The statement "it never was nor will be, since it is now, all together" seems to claim not merely that what is will not *come to exist,* but that it will not exist *at all* in the future" (p. 250). Again, they must think they are able to imagine Parmenides' changeless reality somehow marching, or failing to march, forward into the future. Their complaint misses the likelihood, as argued above, that Parmenides' stand against becoming is fundamental. When Kirk, Raven, and Schofield admit Parmenides probably meant to "ascribe to what is existence in an eternal present not subject to temporal distinctions of any sort," yet they still complain they do not see how he grounded such a conclusion, part of *their* problem might lie in *their* idea of an "eternal present." Parmenides would see it to be a despised "turning back" from their recognizing what is is not subject to "temporal distinctions of any sort."

[18] "I have developed this interpretation in (1985b). It takes as central Plato's remarks in his *Parmenides (141f.* and 156f.), but see also Timaeus (37f.) and *Sophist* (249f. and 258ff.).

[19] In a similar vein, some may try to understand the all-togetherness of Parmenides' One on the model of everything (that has or will happen) being *present to* God in one divine glance. But why think such an idea can give new sense to "eternal"? The temptation involves an equivocation, a shift from a presenting (a representation or intuition) *of* all "eternity" to a presentation (or presence?) that is eternal (always there). There might be a representation of eternity that is itself temporal and not everlasting; given the character of reality, it is harder to think of one that is everlasting. But if God's consciousness of everything is *not* to be merely everlasting, does this not make it atemporal – and more mysterious? In any case, this does not help with Parmenides: if the diverse temporal world is not real, what is the point of supposing it glimpsed by a singularly eternal (in any sense of "eternal") consciousness?

REFERENCES

Cornford, F. M. (1939), *Plato and Parmenides* (New York: Routledge).

Gödel, K. (1959), "A Remark About the Relationship Between Relativity Theory and Idealistic Philosophy," in P. Schilpp (ed.), *Albert Einstein: Philosopher Scientist* (New York: Harper): 557-565.

Grünbaum, A. (1967), *Modern Science and Zeno's Paradoxes* (Middletown: Wesleyan).

Grünbaum, A. (1969), "The Meaning of Time," in N. Rescher (ed.), *Essays in Honor of Carl G. Hempel* (New York: Humanities).

Guthrie, W. K. C. (1965), *A History of Greek Philosophy* (New York: Cambridge).

Heidegger, M. (1961), in R. Manheim (trans.), *An Introduction to Metaphysics* (Garden City: Doubleday).

Heidegger, M. (1972), in J. Stambaugh (trans.), *On Time and Being* (New York: Harper and Row).

Heidegger, M. (1975), "The Anaximander Fragment," in D. F. Krell and F. A. Capuzzi (trans.) *Early Greek Thinking* (San Francisco: Harper and Row).

Hoy, R. C. (1978), "Becoming and Persons," *Philosophical Studies* XXXIV: 269-280.

Hoy, R. C. (1982), "Ambiguities in the Subjective Timing of Experiences Debate," *Philosophy of Science* XLIX: 254-262.
Hoy, R. C. (1985a), "The Given and the Self-Presenting," *Nous* XIX: 347-364.
Hoy, R. C. (1985b), "Plato's Time Skepticism" presented at the American Philosophical Association Eastern Division Meetings.
Husserl, E. (1966), M. Heidegger (ed.), and J. Churchill (trans.) *Phenomenology of Internal Time-Consciousness* (Bloomington: Indiana University Press).
Kant, I. (1965), "Anticipations of Perception" and "Second Analogy," in N. K. Smith (trans.) *Critique of Pure Reasons* (New York: St. Martin's Press).
Kirk, G. S., Raven, J. E., and Schofield, M. (1983), *The Pre-Socratic Philosophers* (New York: Cambridge).
Mellor, D. H. (1981), *Real Time* (New York: Cambridge).
Owen, G. E. L. (1974), "Plato and Parmenides on the Timeless Present," in A. Mourelatos (ed.), *The Pre-Scoratics* (Garden City: Anchor/Doubleday): 271-292.

10
REICHENBACH ON INDETERMINISM AND BECOMING

Iain Martel
University of Colorado at Boulder

On first inspection, there would appear to be a close connection between the thesis of indeterminism and the belief that, while the past and present are real, the future is not real.[1] The claim that the future is indeterminate sits nicely with the claim that the state of the world at the present time (together with the laws of nature) does not determine what the state of the world will be at a later time. This intuitively plausible connection between indeterminism and a dynamic view of the world has been suggested by a number of philosophers, some of whom believed that indeterminism is both a necessary and a sufficient condition for the world to be a dynamic one. Hans Reichenbach is widely taken to have defended this position, and a number of authors – including Michael Tooley (1997; § 1.4), Richard Healey (1981; § 3) and Adolf Grünbaum (1963; § ch. 10) – have criticised Reichenbach for this view. In this paper, I will argue that each of these critics has, in one way or another, seriously misrepresented Reichenbach's views, and that their criticisms are therefore almost entirely misdirected.

I will begin by looking at criticisms offered by Michael Tooley and Richard Healey against the view that there is a logical tie between determinism and the dynamic view of the world.

1. INDETERMINISM, BECOMING, AND VERIFICATIONISM

In his recent book, *Time, Tense, and Causation,* Michael Tooley argues against both the necessity and the sufficiency of indeterminism for a dynamic world. The argument against the sufficiency of indeterminism is simply the observation that a static world can perfectly well contain indeterministic laws. (For example, if laws are just regularities, then an indeterministic static world would be one in which all of the facts exist, and

L. N. Oaklander (ed.), The Importance of Time, 131–141.

some of the regularities among those facts are statistical in nature.) As we will see, I have no objection to this argument, so long as we restrict the notion of truth to that of truth *simpliciter*. However, I believe that if we adopt instead the notion of *truth-at-a-time*, then indeterminism is a sufficient condition for the dynamic view. I shall postpone discussion of this point until later.

Tooley's argument against the necessity of indeterminism for a dynamic world is essentially that indeterminism can only establish an asymmetry between past and future on the assumption that the notions of events becoming actual or coming into existence make sense independently of the truth of indeterminism, and that these notions render the appeal to indeterminism superfluous. It is easy to see how the past could be determined and the future not on the grounds that the past is actual and the future not, but this would beg the question in favour of a dynamic view. In order to draw a link between indeterminism and the dynamic view, we therefore need to find an argument for an asymmetry of determination between past and future that does not begin by invoking ontological asymmetry. Tooley does not see there being any such argument, and so rejects the claim that indeterminism is necessary for a dynamic world.

Richard Healey also rejects any inference from indeterminism to an objective asymmetry between past and future, but he does see that there is at least one initially promising line of argument that would, if successful, yield the argument that Tooley finds lacking. This is the argument from verificationism. The argument starts from the fact that, "If quantum mechanics is true, then there is no way even in principle that we could presently know the truth value of certain statements about the future" (Healey, (1981), p. 112). From this, the verificationist concludes that such sentences, being unverifiable, are meaningless, or have no truth value.[2] It is further argued that there are no such statements about events in the past, and that there is therefore a fundamental ontological distinction between the past and the future.

Healey doesn't think this sort of argument is successful because he thinks the verificationist criterion is so strong that it rules out knowledge of past events as well as future ones, and it does so even in a deterministic world. The main point of Healey's argument is the claim that, even in a deterministic world, a "knower is only able to use knowledge of the past state to acquire knowledge of the future state if he, she or it also knows that the system remains undisturbed by outside influences in the intervening period" (1981, 114). As this knowledge can never be available at the time of prediction, the verificationist is forced to conclude that we can have no knowledge of the future even in a deterministic world, thus severing the link between indeterminism and the unreality of the future. Furthermore, the same argument could plausibly be applied to show that we can have no knowledge of the past either, thus destroying the claimed asymmetry between past and future.

Healey's argument here depends on a very strong interpretation of the verifiability criterion – historically, that of the early logical positivists. This criterion requires that a proposition be capable of conclusive verification if it is to be meaningful. Reichenbach, however, defended a much weaker verifiability criterion, and with this criterion, I will argue, neither step of Healey's argument goes through. On Reichenbach's verifiability criterion, knowledge of the non-existence of interfering influences is not a requirement for a prediction to be justified, or to have a truth-value. On this account, the existence of a truth value for a proposition only depends on the possibility of there being grounds for believing with arbitrary probability (<1) that the statement is true, not on the possibility of certain knowledge. And with this weakened condition, we will see that statements about the future in a deterministic world, and statements about the past, do possess a determinate truth-value, for it is possible to have grounds for believing that such statements are probably true or probably false and this probability can be made arbitrarily high by increasing the precision and scope of the evidential basis. Statements about the future in an indeterminate world, however, will often lack a determinate truth-value, for there will be statements about the future such that it is impossible, in principle, to have evidence beyond a certain probability for the belief that they are true – or, in some cases, that they are even probable.

Let us take a more detailed look at this proposal. We start with the statement of what might be called "weak probabilistic verificationism:"

> WPV: A proposition p is true (false) iff evidence could be obtained that gives a probability for the occurrence (non-occurrence) of the state of affairs denoted by p, and the probability can be increased without limit towards one.[3]

This is roughly the form of verificationism defended by Reichenbach in his epistemological treatise, *Experience and Prediction.* As it stands, however, WPV is open to a number of interpretations, which we must choose between. The first main clarification required concerns limitations on who (what) can obtain evidence, how they may obtain it, and where and when they may obtain it. Concerning the first two of these constraints Reichenbach is quite liberal – all that is required is that some measuring device (be it a machine, a human mind, or any physical object capable of recording information) be able to obtain relevant information in some way.

On the question of where and when evidence must be available for a proposition to count as true, there are a number of available alternatives, the choice between which will have an important bearing on the ontology we are committed to. If we only require that the evidence be obtainable *sometime* and *somewhere*, then the resulting notion of truth will be what Tooley calls "truth *simpliciter*," and it would seem that we would have no choice but to accept a static view of the world - what is true at one time is true at all times, and there is no change in the world. However, if we choose to restrict the domain of evidence that is allowed to count in favour of a proposition's truth to that evidence which would be possible to obtain at a particular time,[4] then we will have a notion of truth-at-a-time, a notion which would appear to leave open the question of whether the world was static or dynamic. The option of allowing only the concept of truth *simpliciter* thus vindicates Tooley's argument that a static world is compatible with indeterminism. However, in his writing on temporal becoming, Reichenbach clearly opts for the notion of truth at a time, holding that what is determinate is what is verifiable as of a particular time.[5]

A second important clarification concerning the interpretation of WPV concerns the scope of the modal term "could" in its expression. What does it mean to say that evidence *could* be obtained in favour of a proposition? It is important that the scope of "could" not be too wide, or past, present and future will all turn out to be real, and it is important that the scope not be too narrow, or much of the past will not count as real. Thus, we cannot interpret "could" as indicating a mere logical possibility, for it is arguably logically possible to obtain evidence for any proposition, whether concerning the past, the present, or the future. Similarly, it would not do to interpret the "could" as "could, given the actual circumstances" –as Reichenbach himself points out, "there exist many past facts which can no longer be known because they have not been registered" (Reichenbach (1952), p. 275). Such an interpretation would lead to the claim that much of the past no longer exists; that many statements concerning the past have no truth value. While some empiricists have been happy with this conclusion, it is not one, which Reichenbach would have accepted.

How, then, are we to interpret the "could" of "could be obtained," given that our interpretation must be neither too wide nor too narrow? The answer is that we must interpret it as, "could, consistent with the laws of nature, have been obtained, if the conditions surrounding the event in question had been suitable."[6] For example, although an electron may, in fact, have passed through a region undetected, it could have been detected, had there been a measuring device in place. We shall see that this interpretation of the modal clause of WPV is sufficient to establish an objective asymmetry between the past and the future.

With these clarifications of WPV in mind, we are now in a position to state a precise version of the verificationist definition of truth-at-a-time, and to see how it applies to facts past and present in deterministic and indeterministic worlds. Here is our definition:

> WPV_t: A proposition p is true-at-t iff were conditions suitable, evidence would be available at t that would give an arbitrarily high probability for the state of affairs denoted by p.

Now, for any event that happens in the world, it is, in principle, possible to record the occurrence of that event by means of a correlation between the occurrence or non-occurrence of that event and different states of some macroscopic system. This possibility seems to be built into the laws of nature. Further, for any macroscopic state of a system, the laws of physics allow that a trace of that state may be preserved over a period of time – at any time later than the occurrence of the initial macro-state, it is possible for there to be a record of the occurrence of that state, where a record is a state, such that the occurrence of that state is highly unlikely unless the state of which it is a record actually occurred.

Putting together these two facts which, again, seem to arise directly from the laws of physics, we see that, for any event in the past (strictly, the absolute past), it is, in principle, possible that there could now exist a record of the occurrence of that event, and the record can determine any arbitrary degree of probability for the prior occurrence of that event. According to WPV_t, all statements referring to events in the past thus have determinate $truth_t$ values – the past is $real_t$.

We may, by very similar reasoning, conclude that, in a deterministic world, the future is also $real_t$. For if all past events are $real_t$, and could be known to have occurred, then it is, in principle, possible to know the exact state of a particular time slice of the universe to arbitrary accuracy. This information, together with the (deterministic) laws of nature, could then be used (by some Laplacian superman) to determine the future state of the universe, again to arbitrary accuracy. Thus, WPV_t entails that all events, past, present and future have determinate truth values at all times in a deterministic world. There can be no becoming in a deterministic world if we accept WPV_t.

What then, of statements about the future in an indeterministic world, as ours appears to be? It is here that Reichenbach's argument (1952, pp. 274-276) concerning the interaction of quantum mechanical indeterminism and the second law of thermodynamics comes into play. According to Reichenbach, Heisenberg's uncertainty principle rules out the possibility of predicting the outcomes of certain future measurements, and does so no matter how well we know the state of the system at the present time. Reichenbach is careful to point out that this uncertainty also applies to retrodictions about past states of the system if all that we know is the state of the system at the present moment. A difference between past and future still exists, however, owing to the possibility of having records of the results of past measurements, but not of future measurements.

It is the second law of thermodynamics, which only licenses inferences to common causes, and not to common effects, which rules out the possibility of having records of future events, and which thus rules out this second source of evidence that we have for past events. Quantum mechanical uncertainty and the second law thus conspire to allow knowledge of the past, but not of the future. There are future facts for which it is, in principle, impossible for us to obtain any evidence that gives even probabilistic justification for a belief in their occurrence. According to WPV_t, such facts thus correctly lack any truth value, and are not $real_t$.

We have thus arrived at the conclusion we sought. We have given a definition of truth, WPV_t, according to which there is an objective ontological difference between the past and the future if and only if the world is indeterministic. This account satisfies Tooley's demand for a rationale for holding that past events are determined but future events not and which does not invoke independent notions of events becoming actual or coming into existence. By switching from the doctrine of strict verificationism to the more sophisticated version of WPV_t, we have also evaded Healey's objections to the verificationist argument.

This argument cannot, of course, be held to have established the connection between indeterminism and the dynamic view of time. For the argument depended crucially on the assumption of verificationism and, as verificationists are a dying breed, such an assumption cannot be made without considerably more defense than could be mounted here. Rather, my goal has been far more modest: I have endeavored to show that, *considered within Reichenbach's epistemological and metaphysical framework*, it cannot be said that "the issue of determinism vs. indeterminism is totally irrelevant to whether becoming is a significant attribute of the time of physical nature independently of human consciousness" (Grünbaum (1963), p. 321). In section two, I shall show that such a claim can only be made by someone who ignores the context of Reichenbach's wider philosophical position, as I believe the author of this claim – Adolf Grünbaum – does.

2. GRÜNBAUM'S OBJECTIONS

Easily the most influential of the critiques of Reichenbach's theory of time is that offered by Adolf Grünbaum in his *Philosophical Problems of Space and Time.* Grünbaum's critique has, indeed, formed the basis for much of the literature on Reichenbach's position, many authors apparently looking no further than Grünbaum's account for their understanding of Reichenbach's.[7] This is unfortunate, for I believe that Grünbaum's account of Reichenbach's position is seriously flawed – at least, I believe that the position Grünbaum criticises is not that of the Reichenbach of 1952, as he claims, but that of the Reichenbach of 1925, which I believe to have been largely contradicted by Reichenbach's later work.[8]

Grünbaum's main criticism of Reichenbach's account comes in two parts. The first part of the argument concerns Reichenbach's claims for the objectivity of "becoming;" it is this part of the argument which I believe most betrays Grünbaum's misinterpretation of Reichenbach's position, circa 1952. Grünbaum argues that Reichenbach has the "avowed aim of providing a physical basis for a 'unique,' transient now and thus for 'becoming,'" and that this aim is not met by Reichenbach's theory because "every 'now,' be it the 'now' of Plato's birth or that of Reichenbach's, always constitutes a divide in Reichenbach's sense between its own recordable past and its unpredictable future" (Grünbaum (1963), p. 322).

I believe that the "aim" that Grünbaum describes was the aim of the Reichenbach of 1925 – it is from here that Grünbaum draws support for this claim – but I do not believe that this aim is consistent with Reichenbach's position as of 1952, and I see no evidence to suggest that Reichenbach thought it was. Indeed, the following passage from Reichenbach's Poincaré lectures of 1952 (this passage, interestingly, is one of the few relevant passages of this work which Grünbaum does not quote) suggests strongly that Reichenbach no longer saw any need for there to be a unique "now" for "becoming" to be an objective feature of the world:

> An act of thought is an event and thus defines a position in time. If our experiences always occur in the framework of a "now," this means that each act of thought defines a reference point. We cannot avoid the "now" because the attempt to avoid it would involve an act of thought and would thus define a "now." A thought without a reference point does not exist, because thought itself defines it. Grammar expresses this fact by the rule that each proposition must contain a verb, that is to say a reflexive sign indicating the time of the event of which one speaks; for the tense of a verb has a reflexive meaning.
>
> Token-reflexiveness applies also to the concepts "determined" and "undetermined".... If we say that the past is determined, or that the future is

> undetermined, it is implied that this is relative to the present situation; it is relative to the "now" that the past is determined and the future is not. These words, and many others, thus have a token-reflexiveness. It is true nevertheless that these words express an objective relation; for it is a physical fact that, if A is the state defined by the act of speaking, then a state preceding A is determined with respect to A, while a state which follows is not. (Reichenbach (1952), p. 277)

It is clear, I think, from this passage that Reichenbach fully acknowledges that "every "now"... constitutes... a divide between its own recordable past and its unpredictable future," but does not believe that this makes "becoming" any less objective. When Reichenbach worked out an account of token-reflexive sentences that succeeded in preserving the objectivity of the relations expressed by such sentences (Reichenbach (1947), § 50), the problem of defining a unique "now" no longer bothered him – objectivity could be preserved without it. Grünbaum's criticism thus addresses a position that Reichenbach had long abandoned; it has nothing to say against Reichenbach's considered opinion.

Although we have seen that the first part of Grünbaum's criticism was misdirected, it might still be thought that Grünbaum was onto something with his criticism; that Reichenbach's mature theory fails to provide an essential element for "becoming" to be fully objective that the earlier theory strove unsuccessfully to provide. This suspicion is perhaps best addressed by a consideration of the second, more important, part of Grünbaum's criticism, a criticism which I believe reflects a true and important difference between Reichenbach's and Grünbaum's positions, a difference which I nevertheless believe that Grünbaum is unaware of.

Grünbaum's objection here rests on what he sees as the confusion of "two quite different things" in Reichenbach's account:

(1) the epistemological precipitation of the actual event properties of future events out of the wider matrix of the possible probabilities allowed by the quantum mechanical probabilities, and

(2) an existential coming into being or becoming actual or real. (Grünbaum (1963), p. 324).

For Grünbaum, the only sense of "becoming" that is meaningful is "no more than the entry of [an event's] effect(s) into the immediate awareness of a sentient organism (man)." This is because, he believes, in either a deterministic or an indeterministic universe, "it is a fact of logic that what will be, will be" (Ibid). Grünbaum thus denies that any sense can be made of the notion of what is actual as of a given time, or of the corresponding semantical notion of what is true as of a time. But we have seen that perfect sense can be made of these notions, for we gave a definition of truth in terms of the knowable, and there is no difficulty in understanding the concept of what is knowable as of a given time. Grünbaum's inability to comprehend the notion of "becoming actual" as an objective feature of the world and his accusation that Reichenbach confuses "epistemic becoming" with "ontological becoming" thus both arise out of the same misunderstanding. Grünbaum cannot understand "becoming actual" because he does not understand the notion of truth at a time; this is because he does not acknowledge Reichenbach's epistemic conception of truth; and this failure is exactly why he thinks Reichenbach is confused in identifying epistemic becoming and ontological becoming. The identification of the "epistemological precipitation" of events out of the future with the "precipitation into existence" of those events is not a confusion according to the epistemic conception of truth – it is simply a matter of definition.

For Reichenbach, "determined" = "determinate" *by definition*. This means that an event's becoming knowable *is* its becoming actual, and Grünbaum's objection is entirely defused. We can also see why the token-reflexive character of "now" does not make the ontological gulf the "now" marks any less objective, for if "determined" is also token-reflexive, then so too is "determinate" – what is determinate is always relative to a "now," and hence in a continual flux. As every event marks a "now," so too every event marks the "actual."

CONCLUSION

In disputes concerning the nature of time, there is a suspicion that the disputants are often simply talking past each other. In discussing criticisms of Reichenbach's views on "becoming" this suspicion seems vindicated. Here, Reichenbach's views are tied to a particular conception of truth, and yet this fact has been largely ignored by the critics. The result is that the criticisms of Reichenbach's view tend to depend on metaphysical and epistemological frameworks antithetical to the one within which Reichenbach was operating, and within which his discussion of becoming is meaningless. The fundamental absurdity of these discussions is perhaps best brought out by consideration of the fact that Reichenbach was a logical empiricist. As such, he was heavily committed to the repudiation of metaphysics (as traditionally conceived), and yet is being credited with defending a heavily metaphysical thesis. The problem that Grünbaum, Healey and Tooley are addressing is one that the logical empiricist is committed to dismissing as a pseudo-problem, not to defending a substantial answer. Only when Reichenbach's discussion of becoming is interpreted in light of his verificationism can we make sense of this apparent inconsistency.

When we interpret Reichenbach's defence of the dynamic view of time in light of his version of the verificationist theory of truth, we find that the apparent heavy metaphysical import of the dynamic view is reduced to a claim about the scope of our epistemic access to the past and future. Furthermore, by looking closer at the theory of truth defended, we can see that for Reichenbach even this weaker conclusion depends on our making a conventional decision to adopt a notion of truth-at-a time and not a notion of truth *simpliciter*. Either notion can be defined within the verificationist theory of truth, yet one yields a dynamic view of time and the other yields a static view. Ultimately, therefore the choice between the dynamic view and the static view is itself a conventional one – Reichenbach simply chose the dynamic view because it retains what he called the "emotive significance" of time.

All of this is obviously a far cry from the discussions of Grünbaum and Tooley, and even Healey, who discusses the relevance of the verificationist theory of truth, does not consider Reichenbach's considerable deviation from the strict verificationism of the early logical positivists. While I am sure that each of these authors would, if pressed, offer numerous criticisms of Reichenbach's theory of truth, and, in turn the theory of time, the fact is that they do not. And this means that their discussions of Reichenbach's views are largely irrelevant. And perhaps we can draw a larger moral from this discussion: could it be that many more of the seemingly intractable disagreements concerning the nature of time are rooted in unrecognized disagreements concerning the basic metaphysical and epistemological framework which underwrites the discussion?

NOTES

[1] In this paper, I shall assume that this is the only dynamic view of time available, ignoring the alternative of presentism, according to which only the present is real. The question at issue is that of whether there is an objective basis for the notion of *Becoming,* the moving *now* that is held to mark an ontological divide between past and future. Presentism, by itself, is of no help in giving sense to the intuitive idea of an asymmetry between past and future, and it is this that indeterminism promises to provide.

[2] The latter alternative here is clearly preferable. Propositions about the future are not meaningless in the way the verificationist would claim propositions about, say, God to be. They will have a truth value, and so are meaningful. We thus hold to a tenseless account of meaning.

[3] On a correspondence theory of truth, the corresponding notion of the actual will be given by the definition: A state of affairs is actual if "x occurs" is true.

[4] To make the account compatible with the Special theory of Relativity, we would also want to restrict the domain to that available at a particular place at that time. The relativistic

version of the account being offered here would equate "past" with "absolute past," "present" either with the "here-and-now" or the "observable now" (i.e. the past light cone), and "future" with anything outside the past light cone. In what follows, I shall for convenience ignore the complications arising from relativity theory, as they do not affect the substance of the discussion.

[5] Each conception of truth appears perfectly intelligible, and there can, for a logical empiricist, be no a priori justification for choosing one over the other. The realist, of course, would be faced with the task of discovering which conception is the "correct" one, but this question does not trouble the logical empiricist. Reichenbach appears to have opted for the concept of truth-at-a-time partly on the (rather Kantian) grounds that it is the only position which allows for the possibility of free will. This justification strikes me as highly dubious, but fortunately it is not required for the theory under development.

[6] This, roughly, is the notion that Reichenbach uses in *Experience and Prediction* when he talks of the "physical probability meaning."

[7] I am thinking here particularly of Storrs McCall's account in his (1976), which demonstrates a significant misunderstanding of Reichenbach's account, arising, it seems, from complete reliance on Grűnbaum's discussion.

[8] Reinchenbach's posthumous book, *The Direction of Time,* explicitly rejects the probabilistic theory of causality offered in Reichenbach (1925), and I believe that much of the rest of the material of that paper was later rejected, or at least considerably modified, by Reichenbach.

REFERENCES

Grünbaum, A. (1963), *Philosophical Problems of Space and Time* (New York: Alfred A. Knopf).

Healey, R. (1981), "Statistical Theories, Quantum Mechanics, and the Directedness of Time," in Healey (ed.), *Reduction, Time and Reality* (Cambridge: Cambridge University Press).

McCall, S. (1976), "Objective Time Flow," *Philosophy of Science* 43: 337-362.

Reichenbach, H. (1925), "The Causal Structure of the World and the Difference Between Past and Future," trans. in Reichenbach (1978): 81-119.

Reichenbach, H. (1938), *Experience and Prediction* (Chicago, IL: University of Chicago Press).

Reichenbach, H. (1947), *Elements of Symbolic Logic* (London: Macmillan).

Reichenbach, H. (1952), "Les Fondements Logiques de la Mécanique des Quanta," trans. as "The Logical Foundations of Quantum Mechanics" in Reichenbach (1978): 237-278.

Reichenbach, H. (1956), *The Direction of Time* (Los Angeles: University of California Press).

Reichenbach, H. (1978), *Selected Writings, 1909-1953* (Dordrecht: Reidel).

Tooley, M. (1997), *Time, Tense, and Causation* (Oxford: Oxford University Press).

11
TIME FLOW

Storrs McCall
McGill University

Majority opinion among scientists and philosophers of science is against time flow or temporal passage as a property of the physical world. Eddington, writing in 1920, gives an eloquent and poetic statement of the scientific position:

"Events do not happen; they are just there, and we come across them. 'The formality of taking place' is merely an indication that the observer has on his voyage of exploration passed into the absolute future of the event in question" (Eddington (1920), p. 51).

The alternative to considering temporal becoming as a physical process is to make it a subjective phenomenon, as Adolf Grünbaum does in putting forward the "mind-dependence" theory.[1] Grünbaum is careful to distinguish between temporal *anisotropy* and becoming. To say that the four-dimensional world is anisotropic along the time axis is to say that there exist processes which take place either normally or invariably in one of the two temporal directions, and rarely if ever in the other. In four-dimensional language, there exist four-dimensional volumes corresponding to these processes which exhibit a characteristic structure or "grain" in the temporal dimension, this grain having a uni-directional pattern. For example, hot and cold water mix to form lukewarm water; lukewarm water almost never "unmixes" into hot and cold components. A match lit on a dark night emits concentric expanding spheres of photons; rarely if ever do concentric shrinking spheres converge on a single point.

But the anisotropy that the world exhibits along the time axis, of which these are examples, does not by itself serve to pick out one of the two temporal directions as THE direction of time. It is arguable that only the property of flow or becoming can do this. If time flows, the direction of time is of necessity the direction in which it flows.

This latter point requires elaboration. If there is such a thing as temporal passage or flow, then it is a matter of logic that the flow is *from* the past *to* the future. It would be absurd to suppose otherwise, and postulate that time might flow from future to past. Hence if there is time flow its existence

L. N. Oaklander (ed.), The Importance of Time, 143–151.

ipso facto (i) divides the time axis into a past end and a future end, and (ii) distinguishes the past-to-future direction as THE direction of time, in opposition to the future-to-past direction.

Neither of these things can mere anisotropy accomplish. If tomorrow we entered into a period in which lukewarm water habitually separated into hot and cold components, or in which entropy habitually decreased, this would not by itself imply that time had changed direction. Hence time flow, if there is such a thing, is logically stronger than temporal anisotropy. None of this goes to show that flow or becoming exists, but only indicates the area in which the truly distinctive character of time resides.

Let us return for a moment to Grünbaum's mind-dependence theory. For Grünbaum, a universe which contained no conscious beings would exhibit no temporal becoming. In such a universe, things would have temporal extension, i.e. would be of a certain age as measured by the difference between their time coordinate in some reference frame and the time coordinate of their origin. But nothing in the strictest sense would "grow older," or "become larger," or change in any way with the passage of time, since in a universe without conscious beings there would be no passage of time. Something X's "becoming larger" in a four-dimensional world means only that later stages of X's world line have broader cross-sections than earlier stages. For Grünbaum, temporal passage or flow is a type of subjective illusion, in the same category as Locke's secondary qualities of colour, taste, and smell.

This is a perfectly consistent philosophical position, one that coheres elegantly with relativity theory. In Minkowski's formulation, special relativity requires replacing the familiar world of three-dimensional objects changing in time by a four-dimensional world in which all "changes" are represented by the interweaving and alteration in cross-section of the world lines of four-dimensional individuals. Thus a given four-dimensional volume may have a three-dimensional cross-section at one time which has the shape of a frog, and another cross-section at an earlier time which has the shape of a tadpole. Change in the four-dimensional world is no more than this.

The four-dimensional world is static not dynamic; it represents change within it, but does not itself change. In the (unbranched) four-dimensional world, time most emphatically does not flow. Hence Grünbaum's theory, which locates time flow wholly within the domain of conscious experience, appears to be ideally suited to the world-view of modern physics.

Although not endorsing time flow, modern physics does not rule it out as a logical impossibility. Its attitude is empirical rather than logical; time flow may exist, but (i) there is no hard experimental evidence for it, and (ii) it plays no role in physical theory. A stronger position is taken by those philosophers who follow McTaggart in claiming that the whole notion of things or events changing their temporal status, from "future" through "present" to "past," is literally absurd.[2] For these philosophers there not only *is* no such thing as time flow; there *could not* be. But, as will be shown, McTaggart's alleged proof of the impossibility of temporal becoming rests upon an error.

The proof runs as follows. The distinctions "past," "present" and "future" apply to events, and they are incompatible. Every event must have one or the other, but no event can have them all. Nevertheless, McTaggart claims, every event *does* have them all. If an event *e* is past (*Pe*), it has been present (*Ne*); if present, it was future (*Fe*). For any event *e* we have *Pe* and *Ne* and *Fe*. But these three are incompatible. To suppose that they are all true is a contradiction.

The obvious reply is that *e* is not *simultaneously* past, present and future, but e.g. *is* present, *was* future, and *will be* past. But according to McTaggart this move to compound tenses will not help, for no event can have all nine of the compound tenses *PP, PN, PF, NP, NN, NF, FP, FN, FF*. In particular, *PP* and *FF* are incompatible. Every attempt to get rid of the incompatibility by moving to more complex compound tenses is met by a further incompatibility at the higher level.

This argument is a sophism. At *no* level do all the tensed attributions simultaneously hold of any event *e*, so that although *PPPe* and *FFFe* are incompatible no one ever would, or could, truly assert more than one of them at a given time. "X is hot" and "X is cold" are incompatible, but no contradiction results if they are not asserted about the same thing at the same time. As will be seen below, there exists an ontological model of time flow which provides a clear proof of the consistency of the concepts "past," "present" and "future," and of the notion of "temporal passage," in the same way that the surface of a sphere provides a clear proof of the consistency of Riemann's elliptic variety of non-Euclidean geometry, or in the way that Tarski's semantics prove the consistency of first-order logic. Hence McTaggart's proof cannot succeed.

McTaggart's views on time, in particular his distinction between the A- and the B-series, have led to an extensive study of temporal language, including (i) the role of tenses and temporal indexicals, and (ii) the question of whether everything that can be said using the latter can also be said using tenseless dated forms. Thus "There will be an explosion" is sometimes analysed as "There *is* (tenselessly) an explosion at time *t*, and *t is* (tenselessly) later than this utterance." The general consensus regarding such de-tensing is that it works with one exception. Tenseless equivalents of almost all tensed utterances can be produced, the exception being tensed statements of the form "It is now six pm, December 29th."[3] This is an interesting and possibly significant result, which I shall not pursue here. Another major development in the study of temporal language has been the emergence of formal tense logic.[4]

None of these results concerning language, however, touches upon the central issue of this paper, which is the ontological status of becoming. Temporal language is important because we want to say things about time without contradiction and without absurdity. But the primary issue is not merely to avoid contradiction, important as this may be. It is whether temporal becoming is a feature of the world, i.e. whether there exists such a thing as the passage of time. If there is, we must of course be careful of the language we use to describe it. But to give linguistic issues priority, and try to draw physical and ontological conclusions from them, is to put the cart before the horse.

So much for preliminaries. I mentioned earlier that for physics, the existence or non-existence of time flow is an empirical matter. McTaggart to the contrary, temporal becoming is not a logical absurdity, and if one day the concept of becoming came to play a role in physical theory analogous to that of "field," or "force," or "virtual particle," or "event," then its existence would be recognized as part of the theoretical framework of science. Nothing rules this out. My object in this paper is to contribute to the process of recognition by showing that a clear and consistent spacetime *model* of time flow is constructible, a model to which the world we live in could conform. The construction of the model will be rigorous and precise, so that any vagueness about becoming (so modelled) which remains in the mind of the reader will be my fault, not the reader's. Besides time flow, many philosophical concepts such as *probability, law of nature, objective chance, quantum non-locality* and *collapse, transworld identity, deliberation* and *decision,* correspond in a natural way to features of the model, the explanatory powers of which are described elsewhere.[5] In what follows I focus on the passage of time, on what it means for events to "happen." Needless to say, happening or becoming will not (in the model) depend in any way on conscious experience, or indeed on the existence of conscious or living beings. Unlike Grünbaum's theory the "becoming" modelled in this paper is entirely objective, was a feature of the world long before life evolved, and will continue to be a feature of it long after living beings have become extinct.

Briefly described, the model is this. We are to think of the world as a huge four-dimensional collection **W** of point-events. **W** includes the Minkowski world of special relativity but it also includes a great deal more: not only the actual events of the world but all those that are physically possible as well, relative to conditions prevailing at the time of the universe's origin. The events in **W** are arranged in subsets called branches (defined below). **W** as a spacetime is distinguished from all other spacetimes, for example the spacetime of general relativity, by virtue of containing pairs of "inaccessible" events. x and y are mutually inaccessible if there is at least one branch that contains x but not y, or y but not x.

In **W**, the passage of time is modelled by progressive branch loss. **W** has a complex densely branching structure which it is almost impossible to visualize, but there are orderly subsets W_f of **W** (the parameter f denoting frame of reference) which have a recognizable tree-like shape. Temporal passage is modelled by progressive branch attrition in each sub-model W_f. Time flow, or branch loss in a frame, will be defined once we succeed in isolating the subset W_f from **W**. But if we wish to change frames and represent time flow in a different frame f', then we must go back to **W** and construct the new subset $W_{f'}$. How exactly this is done will be described in what follows.

Besides **W**, the only other spacetime I know of with the topological feature of inaccessibility is Belnap's "Branching space-time" (BST), to which the formal exposition of this paper owes much.[6] In BST, and in **W**, the fundamental theoretical concept is a binary relation $\leq$ defined on **W**, where $x \leq y$ is read "x is weakly prior to y," or "x weakly precedes y." Strict precedence $x<y$, which corresponds to Einstein's "causal relation," is defined as "$x \leq y$ and $x \neq y$." The relation $\leq$ is a partial ordering on **W**. If we assume that **W** is a topological space, all the other structural concepts of **W** are definable in terms of $\leq$. I shall also sometimes read $x \leq y$ as "y is an upper bound of x."

Df. 1 Two points x and y are COMPARABLE if $x \leq y$ or $y \leq x$; otherwise INCOMPARABLE

Using the notion of comparability we can define the concepts of "timelike path" and "lightlike path" in **W**. We first define a more general concept, that of a one-dimensional subset of **W** known as a "maximal chain."

Df. 2 A MAXIMAL CHAIN in **W** is a maximal set of pairwise comparable points.

(The "maximality" of a chain C consists of C being both "full" of points of **W** and also being maximally long. A maximal chain is never a proper subset of any other chain.)

Df. 3 A maximal chain is LIGHTLIKE or NULL if for any two points x and y on it where $x<y$, every intermediate point z such that $x<z$ and $z<y$ is already a member of the chain.

A lightlike chain in **W** is a set of points along the path of a light ray, i.e. a set capable of being occupied by a light ray, whether or not a ray is actually occupying them. The set of all null chains or paths through any point x defines x's double light cone, one lobe of the cone being "past" and the other "future."

Df. 4 A TIMELIKE PATH is a maximal chain no part of which is lightlike.

In relativity theory, no object with non-zero mass can travel at the speed of light. The "slower" paths which constitute the world lines of bodies with mass are all timelike, not lightlike.

We now proceed to define "branch" and "inaccessibility."

Df. 5 An UPWARD DIRECTED subset of **W** is a set, any two members of which have a common upper bound in the set.

Thus for any x,y in an upward directed set there is a z in the set such that $x \leq z$ and $y \leq z$.

Df. 6 A BRANCH is a maximal upward directed subset of **W**.

Again, the "maximality" of a branch lies in the impossibility of adding any additional points to it which are not already members of the branch. Every point of **W** belongs to at least one branch. Every pair of branches intersects. If the temporal extension of the world is finite, with a last moment, then some points belong to only one branch. If time extends infinitely, then every point belongs to the intersection of two or more branches. **W** branches only forwards in time, not backwards.

We now define accessibility and inaccessibility.

Df. 7 Two points are MUTUALLY ACCESSIBLE if they share a common upper bound, and MUTUALLY INACCESSIBLE if they have no common upper bound.

Df. 8 A set is ACCESSIBLE if it is pairwise accessible, i.e. if any two of its members are mutually accessible.

Alternatively, a set is accessible if it is a subset of some branch of **W**. We now define a "slice," which is an infinitely thin three-dimensional hypersurface of **W**, only one instant thick, no portion of which is either earlier or later than any other portion. A slice in effect defines a total "world-state" of a branch of **W** at an instant.

Df. 9 A SLICE is a maximal accessible set of pairwise incomparable points.

The next concept to be defined is that of a choice point.

Df. 10 x is a CHOICE POINT if there are at least two distinct timelike paths A and B on which x lies, and if those portions of A and B which are each later than x are pairwise inaccessible.

The existence of choice points indicates *that* **W** branches, but does not determine *how* it branches. In fact **W** branches along "corrugated surfaces," which are, like slices, maximal three-dimensional hypersurfaces only an instant thick, but which have everywhere the slope of a light ray. Imagine a discrete set S of pairwise incomparable choice points, each one separated from its neighbors in S by non-choice-points. Imagine also the set of all future light cones of these choice points, each cone intersecting its neighbors. If we discard all portions of these cones which are later than points at which they intersect with neighboring cones, what we are left with is the "corrugated surface" determined by the set of choice points.[7]

Choice points do not occur randomly in **W**, but are distributed in an orderly way along parallel branching timelike paths. With the exception of "main choice points" (MCP's), which are the limit points of upwardly bounded sets of choice points in a path, sets of choice points are discrete. Thus each non-MCP choice point has a unique predecessor in the path and two or more immediate successors, each one on a different branch. Non-MCP choice points are separated by extremely small temporal intervals. On any given path, main choice points are separated by intervals of constant length At of the order of 10^{24} sec., this distribution being relevant to the definition of objective probability values in W.

We now define the parallel sets of corrugated surfaces along which the well-behaved subsets W_f of **W** branch. Every frame *f* defines a *foliation* of **W**, that is to say a partition of **W** into equal-time hyperplanes with respect to the frame *f*. These hyperplanes are separated into distinct equal-time slices on different branches, it being understood that the partitioning of **W** cuts across branches. Consider an equal-time hyperplane H_f, which cuts across many branches and has sections in each branch. The set of choice points on or immediately above H_f determines a non-connected set of corrugated surfaces C_f along which **W** branches. The complete set of surfaces along which W_f branches is a subset of the set of all corrugated surfaces parallel to C_f. Similarly, in a different frame f', a hyperplane $H_{f'}$ defines the complete set of corrugated surfaces along which $W_{f'}$ branches.

For each frame f, a submodel W_f of **W** has now been defined. These submodels have a treelike structure, with branching towards the future and no branching towards the past. *Branch attrition* in each W_f consists in the progressive "pruning" or loss of branches, it being always the lowest branches which disappear. This progressive branch loss is what constitutes, in the branched model, the flow of time.

Suppose for example that the lowest branches in submodel W_f split apart along the corrugated surface C_f at time t in frame f. Then at time t all but one of these lowest branches vanish, leaving the remaining branch to form part of the trunk (the "past") of W_f. If the next surface along which W_f splits is located at time t' of frame f, then all but one of *those* branches are lost at t', and so on. This process of progressive branch loss takes place in every submodel W_f, and makes for objective, non-mind-dependent time flow in each corresponding frame of reference f.

To sum up, there exists in the branched model a way of defining time flow that is both mind-independent and precise. The proposed definition is one which gives becoming or temporal passage a place within the theoretical structure of physics. Conceived of as branch attrition within the overall spacetime structure of the universe, time flow emerges as a physical process unlike any other physical process, but nonetheless one that characterizes the physical world in a real and objective way.

NOTES

[1] Grünbaum (1963, pp. 324-325).

[2] McTaggart (1908), (1927, ch. 23); Dummett (1960); Mellor (1981); Horwich (1987, pp. 18-25).

[3] Smith (1994).

[4] Prior (1967); Øhrstrøm and Hasle (1995).

[5] McCall (1994), (1995a), (1995b).

[6] Belnap (1992). BST, however, lacks the feature of branch attrition.

[7] A more formal definition of corrugated surfaces is given in McCall (2000).

REFERENCES

Belnap, N. (1992), "Branching Space-Time," *Synthese* 92: 385-434.

Dummett, M. (1960), "A Defence of McTaggart's Proof of the Unreality of Time," *The Philosophical Review*, 69: 497-504.

Eddington, A.S. (1920), *Space, Time and Gravitation* (Cambridge: Cambridge University Press).

Grünbaum, A. (1963), *Philosophical Problems of Space and Time* (New York: Knopf).

Horwich, P. (1987), *Asymmetries in Time: Problems in the Philosophy of Science* (Cambridge, MA: MIT Press).

McCall, S. (1994), *A Model of the Universe: Space-Time, Probability, and Decision* (Oxford: Oxford University Press).

McCall, S. (1995a), "Time Flow, Non-Locality, and Measurement in Quantum Mechanics," in Savitt (ed.), *Time's Arrows Today* (Cambridge: Cambridge University Press): 155-172.

McCall, S. (1995b), "A Branched Interpretation of Quantum Mechanics Which Differs from Everett's," in Marion and Cohen (eds.), *Quebec Studies in the Philosophy of Science* (Dordrecht: Kluwer): 229-241.

McCall, S. (2000), "Towards a Lorentz-Invariant Theory of Collapse," in Duffy and Wegener (eds.), *Recent Advances in Relativity Theory*, (papers presented at a conference at the Imperial College, London): 111-122.

McTaggart, J. M. E. (1908), "The Unreality of Time," *Mind* 17: 457-474.

McTaggart, J. M. E. (1927), *The Nature of Existence*, vol.II (Cambridge: Cambridge University Press).

Mellor, D. H. (1981), *Real Time*, (Cambridge: Cambridge University Press).

Øhrstrøm, P. and Hasle, P. (1995), *Temporal Logic: From Ancient Ideas to Artificial Intelligence* (Dordrecht: Kluwer).

Prior, A. N. (1967), *Past, Present and Future*, (Oxford: Oxford University Press).

Smith, Q. (1994), "The Truth Conditions of Tensed Sentences," in Oaklander and Smith (eds.), *The New Theory of Time* (New Haven: Yale University Press): 69-76.

12

THE INCOMPATIBILITY OF STR AND THE TENSED THEORY OF TIME

Quentin Smith
Western Michigan University

Many philosophers think that the Special Theory of Relativity is compatible with the tensed or A-theory of time ("mind-independent temporal becoming"). They have argued that the future can be indeterminate or future relative to a reference frame. However, "relative futurity" is not a tensed or A-determination, but a relation that obtains in a B-series in a universe in which the tenseless theory of time (temporal features consist only of earlier than, later than and simultaneity) is true. I shall argue this point in this paper.

1. RELATIONAL PRESENTNESS

Let us suppose, as we need to in order to have mind-independent temporal becoming, that presentness is not observer dependent. This supposition leads to the following conditionals: If presentness is a relational property, then this theory is compatible with STR but inconsistent with the tensed theory of time (the theory of objective time flow). But if presentness is a monadic property, the theory of time is incompatible with STR but consistent with the A-theory of time (the theory of mind-independent temporal becoming).

If "being present" expresses relational properties, e.g., present relatively to frame F1, being present relatively to frame F2, etc., then presentness is not an A-property, constituting temporal becoming, but is a B-relation in a becomingless universe. We may say "E1 is present relative to frame F2," but this is a tenseless statement; it means only that E1 is simultaneous with frame F2.

To explore this idea further, we need more precise terminology, since "presentness is frame relative" is at best misleading and at worst contradictory. The relational property in question is being present relative to an event on a world line; it is not the property of being present relative to a world line. A worldline is a maximal chain that is a timelike path; it consists of causally related events which are present successively, so if something is present relative to the worldline, it would be simultaneous with a sequence of events none of

L. N. Oaklander (ed.), *The Importance of Time*, 153–156.

which are simultaneous with each other. This, of course, is a contradiction. Since a reference frame includes the successive events on this world-line, we should not talk of being present relative to a reference frame. The relational property of presentness is instead the property of being present relatively to an event E on a worldline W.

However, this relational property is not an A-property but a B-property. For each of the successive events E1, E2, E3 on any given world-line W, there is some other event B that has the property of being present relatively to E1 (or E2, etc.) on W. Each event, such as E1, exists on a plane of simultaneous events, such that each event on this plane is present relatively to E1. For example, event B that is on the plane of simultaneity P1 that runs through E1 is present relatively to E1 on worldline W. This shows that the phrase "presently relatively to the event E on worldline W" expresses a B-property – a relation of simultaneity – not an A-property. It is true at every time that D is present relatively to E1, that C is present relatively to E2, and so on. This is not a universe in which there is temporal becoming.

2. MONADIC PRESENTNESS

Let us pursue the idea that presentness is a monadic property. We can then say that event E1 on worldline W is present, and that any event B that is present relatively to E1 is presently occurring. This is a theory of objective time flow. But our theory is no longer consistent with STR, for if E1 on worldline W1 is present, then there is a privileged reference frame. Presentness, if a monadic property, is not the relativistic sort of property required by STR; it is not being present *relatively* to something, but being present *absolutely*. If E1 is present, no event that is not on the plane of simultaneity of E1 on W1 is present. Consider some event E0 on a non-parallel worldline W0 that is earlier than E1. If E1 is present, E0 has the monadic property of pastness. It can be true that E0 *is present relatively to* some event on some other world line W0. But it cannot be true that E0 *is present*. If it were true that E0 is present, then we would have the implicit contradiction: "E1 is present, E0 is present, E0 is earlier than E1." therefore that it is false that E1 and E0 are both present. What is true is that E0 is present relatively to some event on the worldline W0. But this is not an A-property, an objective time-flow property. Rather, it is a B-relation, the relation of E0 being simultaneous with some event on worldline W0.

If we are to have objective time flow, the worldline W must belong to a privileged reference frame and STR must be false. A critic might respond that if we change reference frames and represent time flow in a different frame, it will be the case that E0 is present. However, if there is objective time flow, we cannot change reference frames in this manner, either physically or in thought. We are on the worldline W1 and E1 is present and E0 is past. Can we imagine that we are on W0 and that E0 is present? We can do this only in the sense that we can entertain what we know to be a false proposition, that *we are on W0 and E0 is present*, and pretend that it is true. But this remains a pretence; the proposition that *E0 is present* is false, and we cannot switch reference frames in an ontologically pertinent sense, which would involve knowing *E0 is present* is true or at least it being the case that *E0 is present.* Once we assume that E1 is present, we have to conclude that E0 is not present. If we want to say with truth that both E1 and E0 are present, we have to talk of their being present relatively to events on different world lines. We can "change frames" in the sense that we know that E1 is present relatively to one event on one worldline and E0 is present relatively to an event on another world line, but we are now talking about relational presentness, which is just the B-relation of simultaneity.

3. MIND-INDEPENDENT TEMPORAL BECOMING IN THE GENERAL THEORY OF RELATIVITY

If there is objective time flow, it is logically necessary that there is a privileged reference frame. Objective time flow and STR are not compatible. But it is compatible with GTR. STR is the vacuum solution to the field equations of GTR; it is the solution for a universe with no matter. It follows immediately that STR is not the solution of the field equations that describes our universe, and thus that the participants in the debate about whether or not "STR is compatible with the tensed theory of time" are working with a false solution to the field equations. The solutions of the field equations accepted by contemporary cosmologists are Friedmann's solutions, which describe an expanding universe that contains homogenous and isotropically distributed matter. (I here ignore all the qualifications required by quantum mechanics.) The Freidmann solutions admit of an objective time flow which consists in the successive becoming present of the surfaces of simultaneity that coincide with the surfaces of homogeneity. Relative to these surfaces, the universe expands (or contracts) uniformly, and matter appears uniformly distributed on a large scale.

In defining objective time flow in a Friedmann universe, we need only the familiar notion of a maximal, spacelike hypersurface. There is a natural foliation of a Friedmann manifold by hypersurfaces of homogeneity and this foliation gives us our privileged reference frame. This is the frame of the worldlines whose surfaces of simultaneity coincide with the surfaces of homogeneity.

Relative to these worldlines, the matter of the universe is distributed homogeneously and isotropically on a large scale, and the universe is expanding uniformly. Objective time flow consists in the successive becoming present of different surfaces of homogeneity.

Part Three

THE PHENOMENOLOGY OF TIME

13
EXPERIENCING THE FUTURE: KANTIAN THOUGHTS ON HUSSERL

Eric M. Rubenstein
Indiana University of Pennsylvania

Attempting to understand our experience of time we confront two images. On the one hand, our experience is depicted as awareness of the present which itself is but an instantaneous, point-like event, one which is forever eluding our grasp.

Time as ordinarily conceived is sundered into separate moments which are perpetually passing away. The past is forever dead and gone, the future is non-existent and uncertain, and the present seems, at most, a bare knife-edge existence separating these two unrealities.[1]

In fact the only time that can be called present is an instant, if we can conceive of such, that cannot be divided into even the most minute fractions, and a point of time as small as this passes so rapidly from the future to the past that its duration is without length.[2]

Alternatively, according to the Phenomenological tradition, our experience is of a temporal spread – a temporal field that stretches well beyond an instantaneous "now." Husserl gives voice to this with talk of a "temporal horizon," for instance, while Merleau-Ponty characterizes this experience in terms of a "field of presence."[3]

> The world now present to me, in every waking "now" obviously so, has its temporal horizon, infinite in both directions, its known and unknown, its intimately alive and its unalive past and future. (Husserl, (1990), p. 92)

It is in my "field of presence" in the widest sense – this moment that I spend working, with, behind it the horizon of the day that has elapsed, and, in front of it, the evening and the night – that I make contact with time … (Merleau-Ponty (1962), p. 416).

Moreover, says that tradition, not only do we experience temporal spreads, not points, our experiences are alive with times other than the present. Past and future times are themselves part of our experience of time.

L. N. Oaklander (ed.), The Importance of Time, 159–169.

Truly ... it pertains to the essence of the intuition of time that in every point of its duration ... it is consciousness of *what has just been* and not mere consciousness of the now-point of the objective thing appearing as having duration (Husserl (1966), p. 53-54).

Everything, therefore, causes me to revert to the field of presence as the primary experience in which time and its dimensions make their appearance unalloyed .... It is here that we see a future sliding into the present and on into past. Nor are these three dimensions given to us through discreet acts ... I do not think of the evening to come and its consequences, and yet it "is there," like the back of a house of which I can see only the facade ... (Merleau-Ponty (1962), p. 416).

All the same, understanding is in every case a Present which "is in the process of having been." All the same, one's state-of-mind temporalizes itself as future which is "making present." And all the same, the Present "leaps away" from a future that is in the process of having been, or else it is held on to by such a future (Heidegger (1962), p. 68).

Phenomenologists thus offer a tantalizing picture of our temporal experience – that contrary to what might be called the "instantaneous-present" tradition, our experience of time is properly seen as one of a temporal spread, one that includes both the past and the future. I aim in this essay to make sense of such a picture, paying particular attention to the claim that our experience of a temporal spread somehow involves the future. After all, *prima facie*, the future does not exist, and it is thereby rather mysterious how we could experience it.

As we will see, focusing on the possibility of a non-existent future entering our experiences parallels what has come to be known as the problem of the specious present. The puzzle I wish to examine is the future-oriented correlate of the specious present, which is itself concerned with the presence of the past in experience. I will argue that accounts of such temporal spreads, ones which include elements of the past and future, can make sense only if we add a Kantian grounding to the phenomenologist's depiction. Namely, we must add the Kantian thesis that time is an *a priori* representation.

1. THE FUTURE IN RECOLLECTION

As an entry point to our discussion I wish to begin with an interesting case. Husserl describes the sort of experience I have in mind.

> Recollection is not expectation; its horizon, which is a posited one, is, however, oriented on the future, that is, the future of the recollected. As the

> recollective process advances, this horizon is continually opened up anew and becomes richer and more vivid. In view of this, the horizon is filled with recollected events which are always new. Events which formerly were only foreshadowed are now quasi-present, seemingly in the mode of the embodied present.[4] (1966, p. 76)

Here we are to focus on the recollection of an experience which itself was temporally extended. As we recall the beginning of that experience we anticipate the remaining temporal portions of that experience. And though Husserl does not provide an example of this sort of experience I believe an apt one would be the recalling of a familiar melody. Thus in rehearsing a melody we anticipate the notes that are to come *after* the ones that we are presently entertaining (or, singing to one's self we might say). Our experience of recollection of that melody is not an experience of temporally discrete moments, where future portions of the melody arise *ex nihilo* as we move along Lovejoy's "bare knife-edge of existence." Rather, even while we are rehearsing a particular note, that experience is permeated with what is to come – those notes which comprise the future portion of the melody. Indeed, we might say that only because we are so aware of those future portions that when it is time to sing *them* that we know how to.

Thus while we are recollecting, and thereby making contact, in some sense, with the past, we are at the same time, in some sense, making contact with future times. Admittedly, this is not contact with the future in a fully rich sense. Rather, it is the contact with the future of a past episode. Even still, this provides a useful starting point for further reflections.

The question I wish to address now is whether we can simply abstract from this case of protentions in recollection to protentions *simpliciter*. Put somewhat differently, is our experience of the future, the kind present in our experience of a temporal spread, based on or derived from this sort of case? Is our representation or experience of the future based on protentions in recollection?

Ultimately I wish to argue that this cannot be the whole story. In short, we must presuppose an experience or representation of the future to even make *sense* of *protentions in recollection*. As we shall see, this is in effect Kant's thesis that time is an *a priori* representation, one which cannot be derived from experiences of temporal relations.

2. THE PAST AND SPECIOUS PRESENT

As a step towards arguing that the future must be presupposed to even make sense of protentions in recollection, let us begin by considering what is required for a representation of the past itself. Here we should consult

Husserl, whose detailed discussions have been so highly influential in articulating our experiences of time. And though Husserl distances himself from explicit talk of the "specious present," rejecting the psychological and empirical elements present in discussions under that title, I take him to be chasing essentially the same phenomenon. Namely, how can an act of awareness which is instantaneous, contain as its object of awareness something which is extended in time? Put differently, but to the same point, how can there be experiences of a duration or succession of events given that certain components of that duration are no longer existent during that *experience* of the duration.

Husserl describes our experience of such duration or objective succession as follows.

> The sound is given; that is, I am conscious of it as now, and I am so conscious of it "as long as" I am conscious of any of its phases as now. But if any temporal phase (corresponding to a temporal point of the duration of the sound) is an actual now (with the exception of the beginning point), then I am conscious of a continuity of phases as "before," and I am conscious of the whole interval of the temporal duration from the beginning-point to the now point as an expired duration (1966, pp. 53-54).

His analysis of how this is possible is explained as follows:

> That several successive tones yield a melody is possible only in this way, that the succession of psychical processes are united "forthwith" in a common structure. They are in consciousness one after the other, but they fall within one and the same common act. We do not hear the sounds all at once, as it were, and we do not hear the melody by virtue of the circumstance that the earlier tones endure with the last. Rather, the tones build up a successive unity with a common effect, the form of apprehension ... Accordingly, there is a perception of temporally successive unities just as of coexisting ones. (1966, p. 41)

> The actual tonal now is constantly changed into something that has been; constantly, an ever fresh tonal now, which passes over into modification, peels off. However, when the tonal now, the primal impression, passes over into retention, this retention is itself again a now, an actual existent. While it itself is actual (but not the actual sound), it is the retention of a sound *that has been.* (1966, pp. 50-51, my emphasis)

In other words, in listening to a melody, those portions which have already been played are no more. Yet if we are to be aware of their existence, they have to be somehow, somewhere, in mind, if the music that we hear now is to form part of a melody. Were those earlier portions gone completely, the sounds we are hearing now would mean nothing. They would be individual, detached notes, not portions of a melody. Therefore, that which is past has to be present in some manner in order for our experience to be of a temporal duration.

Husserl aims then to explain the possibility of such experiences of a temporal duration by describing how those items which are, in some sense, no more, can still be experienced as portions of such a duration. Husserl gives us the following picture to help.

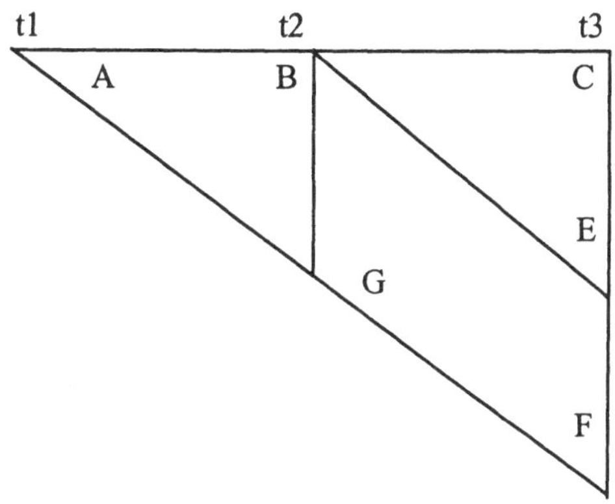

Figure 1

On this account, the horizontal line A-C represents the series of notes experienced during a certain interval. Thus at t1, A is perceived, at t2 B is perceived, and then at t3 C is perceived. The question is how, at t3, can the subject be aware of a *series* of notes having been experienced? How, can the subject be aware of a melody, given that at t3 the only sound present is C?

Husserl's answer, in short, is that when A was experienced at t1, it left its mark on the perceiver, such that it was retained in consciousness. The diagonal line A-F represents this retention; likewise for diagonal B-E, which is the retention of the experience of B at t2. Thus at t3, those original experiences are still present to mind, present as retentions of the original experience. In other words, at t3, though the perceiver is presented with a temporally simultaneous array of sounds, namely the vertical line C-F, the perceiver responds with a representation of a temporal sequence. The perceiver, that is, is aware of a melody, though earlier portions of that actual melody are long gone. Again, though the act of awareness at t3 may be instantaneous, the object of awareness is temporally extended.

Now this analysis has been subjected to countless discussions, criticism, elucidations, etc. I do not wish to enter the fray here, either in point of

Husserlian exegesis, nor of the merits of the account in total.[5] Rather, I simply wish to point out what strikes me as an important yet missing ingredient in this account. Moreover, this missing ingredient will advance our discussions towards the possibility of representing the future.

Through the elaborate mechanism of retentions Husserl describes, the "sinking down" of retentions, the perceiver comes to respond to what is in fact a temporally simultaneous array with a judgment about a temporal succession. What is lacking, however, in short, is how such an array *can* give rise to such a temporally oriented judgment.

Consider the situation at t3 in the above diagram. At that time the perceiver has before consciousness, so to speak, a number of representations. And let us grant that this manifold of representations are unified into a single act of consciousness. That, recall, was what interested Husserl – how elements which are part of a duration, and thus past, can be part of an instantaneous act of awareness. At t3 then the perceiver has this unified array as the content of its awareness. But where and how does the *temporal* element enter into this awareness? What makes it the case that this array, unified as it may be, is experienced as a *temporal* array, rather than a non-temporal one? How does time enter the picture here at all?

Husserl, after all, began with a succession of representations. That is what happens between t1 and t3, namely, that we are affected by a sequence of notes. Via his account of retentions and their "sinking down," Husserl finds a way to get all of those representations into a single one, as happens at t3. And though that is helpful, it is not enough. For once those various representations, retentions, are combined into a single representation, what allows for the various aspects of that representation to have a *temporal* component? Where does the "distant past," "near past," "just past" element come from such that that single representation is of a *temporal* succession? We have retained perhaps a series, but what makes it an awareness of a *temporal* series?[6] That is the key question, and one for which Husserl does not have an answer. In short, Husserl simply seems to helped himself to the temporal element that is needed for a complete account of temporal awareness.[7]

We need then to supplement Husserl's account then. The missing ingredient, I propose, is provided by Kant. In short, where Husserl goes wrong is that there is nothing temporally tagged, as it were, in the retentions which gives a temporal element to the representation at t3. Husserl gives us a single representation of a series, but not of a *temporal* one. What is needed is some way to get time or a temporal element into the picture. Without such an element, there would be no temporal aspect in the representations Husserl speaks of. Time is thereby required, but apparently cannot be derived from the unified representation given at t3. For as we saw, that representation lacks the temporal properties we need. If time is required, but cannot be derived, it must be presupposed as already present. Time is required then as an *a priori* representation. And that of course is one of Kant's deepest points in the Transcendental Aesthetic.

> For simultaneity or succession would not even enter our perception if the presentation of time did not underlie them *a priori*. Only on the presupposition of this presentation can we present this and that as being at one and the same time (simultaneously) or at different times (sequentially).[8]

The way to get such a temporal element into our representations of a duration (and thus of past occurrences), then, is not by trying to derive it from the sequence of retentions, nor from the instantaneous array, but rather, to presuppose time at the outset. And as we shall see, as for the past, so too for the future. The only way to make possible the representation of the future is to presuppose it at the outset.

3. DERIVING THE FUTURE

We have seen that we cannot make sense of a representation of the past *as past* unless time is somehow built into such representations, that is, unless time is presupposed as an *a priori* representation. Before turning to those arguments we had wondered whether the future could be recovered from a representation of the past. More specifically, we inquired into the possibility of deriving a representation of the future from protentions in recollection. From there we turned to what is necessary to represent the past at all, regardless of whether such representations themselves contained a future-looking element. Having seen that such representations of the past themselves presuppose a representation of time, we can now begin to explore what is required to represent the future.

Let us begin with straightforward attempt. Here I have in mind attempts to derive representations of the future from feelings or cognitive states such as anticipation or expectation we may have at a given, present moment. I submit, however, that this approach must fail, for feelings as anticipation or expectation can offer a ground for representations of the future only if we already presuppose a representation of time (namely of future times) in such feelings. Such feelings or states cannot by themselves serve as an account of the representation of the future.

What after all, is a feeling of anticipation? It is a future-looking state; one of waiting for an event to unfold or be given from the future. If so, then anticipation or expectation already contains a future-looking element, already contains a representation (even if just in a thin sense) of the future. That said, we cannot use such states to make sense of or derive a representation of the future – for such future aspects are already built right into those states.

This no doubt raises the temptation to simply equate representations of the future with such feelings or states of anticipation and expectation. That is, one could contend that whatever sense there is to the representation of future times simply amounts to the feelings of anticipation we have, our expectations, etc. However, simply having such a feeling of anticipation would not itself count for us as a representation of the future unless we already knew in some way what it would *be* to represent the future. We must, in other words, already have some idea of what counts as being *future*, if we are to be able to glean from experiences of expectation the sort of future-looking element we are seeking.

Having rejected the straightforward model attempt to make sense of representation of the future with such states as anticipation or expectation, we can now return to the question of whether having gained a representation of the past, as made possible by an *a priori* representation of time, we are in a position to abstract or derive representations of the future.

There are two ways to try this. 1) We might simply try to abstract from experiences of the past and the relation of past *pasts* to past *presents* in order to make sense of the future. Alternatively, 2) we might try to simply generalize from experiences of protentions in recollection, those interesting experiences with which we began our discussions. We shall see, however, that such strategies themselves fail, and for precisely the same reasons that Husserl's attempts to derive representations of the past failed – such representations of the past are possible only if time is represented *a priori*. So too for representations of the future. They are possible only if future times are represented *a priori*. Thus we cannot recover such representations from experiences of the past unless somehow built into those experiences themselves are representations of the future.

To see this, and to exhibit again the Kantian sentiment that time must be given *a priori*, imagine the following sequence of representations:

(1) At t1, x is represented; at t2 y is represented; at t3 z is represented.

With such a sequence of representations on their own, there would be nothing which makes it the case that such a series is represented as a temporal succession. We are, as of yet, that is, lacking the resources which makes it possible to represent (1) as:

(2) x was *before* y which was *before* z, which is now *present.*

Recalling our discussions of Husserl, even if we find a way to unify those various representations, making x, y, and z part of a single representation, there must still be something else added to the story to make it the case that x, y, and z are seen as temporally related.

The various elements not only have to be unified into one act of awareness (which Husserl does accomplish), they must as well be so unified such that each element of that one representation is somehow represented as being temporally related to the other elements (which Husserl does not accomplish). With this in mind let us turn to the future.

Making use of the above example, let us consider y to be a recently past event, and z a present one. As we move from t3 to some later time we may notice that y itself moves into the more remote past, and that z, which *was* present, itself becomes past. From that we might infer that as y was followed by z, and that z itself is now past, z itself will be followed by some *new* event. What's more, the ensuing gap between z and that event which *will* follow is what provides for our sense of the future. In other words, as we have experienced the succession of events, we expect that one that was just present will be followed by one in the *future*, given that in the past, remote events were followed by not so remote, but still past events. This, in essence, would be the attempt to abstract a representation of the future from experiences of the past.

The problem with this attempt, however, is just the problem we saw in trying to make sense of Husserl's account of the specious present. There we saw that a sequence of representations would mean nothing to us unless somehow they were, first, unified into a single act, and what's more, unified in such a way that components of the now unified representation were represented as being temporally related to the other. So too with the representation of the future. Only with such a representation of future elements already presupposed would we be able to speak of the future.

Again, simply holding those various representations themselves together, simply unifying them, is itself not enough for a representation of a temporal succession or series. Unifying them would give us a representation of a series, but it would not be one of a temporal series unless we somehow know that those elements of the single, now unified representation, are themselves temporal elements. And this is possible only if time is already built into our representations – only if time is presupposed and given *a priori*. Those items which occurred in the past could take on a significance for us, therefore, of a future element, only if we already presupposed a representation of the future. We cannot make sense of representations of the future simply by abstracting from experiences of succession.

Having rejected (1), we can now consider strategy (2), which aims to gain a representation of the future by recourse to protentions in recollection-experiences of recall in which there is a future-directed element. However, as we shall see, even making sense of protentions in recollection requires a presupposition of future oriented representations. That is, protentions in recollection are not the innocuous starting points they may have seemed – their very possibility requires an *a priori* representation of time, including a representation of future times.[9]

The point is easily seen, given the above. For in recalling a melody, that we have expectations of what is to come could signal for us a future-oriented state, could be an indication of what a future-oriented state is, only if we already know what it would be to have a representation of the future. Those feelings of anticipation would be conceptualized as future looking only if, in some sense, we already understood what it is for something to *be* future. Thus while protentions in recollection may make for interesting phenomenology, they can play a role in our ordinary experience of the future only if we have presupposed an *a priori* representation of time, namely of future times, in such experiences of protention in recollection. And that of course is just the same Kantian point again.

For simultaneity or succession would not even enter our perception if the presentation of time did not underlie them a priori. Only on the presupposition of this presentation can we present this and that as being at one and the same time (simultaneously) or at different times (sequentially).[10]

NOTES

[1] Lovejoy (1961), p. 75.
[2] *Augustine* (1963), p. 15.
[3] Continental philosophy does not have a monopoly on such sentiments, though. James too gives voice to this notice of a temporal spread. Past and future...conceptually separated by the cut to which we give the name of the present...are to some extent, however, brief, co-present with each other throughout experience. The literally present moment is a purely verbal supposition, not a position; the only present ever realized concretely being the passing moment in which the dying rearward of time and its dawning future forever mix their lights. *A Pluralistic Universe,* p. 254.
[4] Husserl (1996, p. 76) calls this sort of future presence, "protentions in recollection," where protentions are those cognitive states which are future directed or forward looking. I will adopt his terminology where useful.
[5] Miller's *Husserl, Perception and Temporal Awareness* (1984) presents an excellent discussion of these and other issues.

[6] Another way to put the point is to simply ask how the verical line C-F *means, stands for,* or *represents* a temporal series. There is nothing intrinsic to the picture that makes it so. Saying that the perceiver responds to that ordering with a temporal representation is to only gesture towards a solution.

[7] This of course is related to the Kantian point that a succession of representations is not sufficient for a representation of a succession. I can indeed say: My presentations follow one another. But that means only that we are conscious of them as being in a time sequence – in accordance, i.e. with the form of inner sense. Time is not, on that account, something in itself, nor is it a determination attaching to things objectively. Kant, *Critique of Pure Reason,* A38=B55. For a contemporary discussion of this point, see Sellars, *Science and Metaphysics* (1968), Appendix.

Rosenberg's *The Thinking Self* (1986) also takes Husserl to task. Without going into the details here, Rosenberg finds Husserl's reliance on retentions to be inadequate. Instead, Rosenberg proposes that previous representations enter into our unified, instantaneous awareness of a succession as ones of which we are aware that we are, or were, aware of; as items of so called meta-awareness (1986, p. 231). But this account falls prey to the same worries as Husserl's. For again, it is by no means clear what a series of awarenesses, meta-awarenesses, meta-meta-awarenesses, etc. have to do with the representation of a *temporal* series.

[8] Kant, *Critique of Pure Reason,* (A30=B46).

[9] Unfortunately, Husserl says very little about protentions beyond that which we have already made use of.

[10] Kant, *Critique of Pure Reason,* (A30=B46).

REFERENCES

Augustine, St. (1963), in R. Warner (trans.), *The Confessions* (New York: The New American Library).

Heidegger, M. (1962), in J. Macquarrie and E. Robinson (trans.), *Being and Time* (New York: Harper and Row).

Heidegger, M. (1962), in J. Churchill (trans.), *Kant and the Problem of Metaphysics* (Bloomington: Indiana University Press).

Hume, D. (1968), *A Treatise of Human Nature* (Oxford: Oxford University Press).

Husserl, E. (1990), *Ideas Pertaining to a Pure Phenomenology and to a Phenomenological Philosophy* (Dordrecht: Kluwer).

Husserl, E. (1966), in J. Churchill (trans.), *The Phenomenology of Internal Time Consciousness* (Bloomington: Indiana University Press).

James, W. (1943), *A Pluralistic Universe* (New York: Longmans Green).

Lovejoy, Arthur (1961), *The Reason, the Understanding, and Time* (Baltimore: Johns Hopkins Press).

Kant, I. (1996), in W. S. Pluhar (trans.), *Critique of Pure Reason* (Indianapolis: Hackett).

Merleau-Ponty, M. (1962), *Phenomenology of Perception* (London: Routledge and Kegan Paul).

Miller I. (1984), *Husserl, Perception and Temporal Awareness* (Cambridge: MIT Press).

Rosenberg, J. (1986), *The Thinking Self* (Philadelphia: Temple University Press).

Sellars, W. (1968), *Science and Metaphysics: Variations on Kantian Themes* (London: Routledge and Kegan Paul).

14
THE THEORETICAL CHARACTER OF HUSSERL'S ACCOUNT OF TIME CONSCIOUSNESS (COMMENTS ON PROFESSOR RUBENSTEIN'S PAPER)

Ronald C. Hoy
California University of Pennsylvania

Professor Rubenstein is wrestling with an important and fertile problem: how can the temporality of consciousness be understood in terms of strictly present states of consciousness. He is right to turn to Husserl, to wonder whether Husserl's account is a good one, and to see some connection with Kant's view of time. I will suggest, however, that Husserl would say that he solved the problem Rubenstein points to, and, rather than needing Kant's help, Husserl would say that he has gone beyond Kant. My interpretation of Husserl's theory varies somewhat from Rubenstein's, but my intention is complimentary. We have more to learn from this problem and Husserl's theory.

I.

Professor Rubenstein sets up the problem of time consciousness against the background of a distinction between two images of time, one according to which the present is instantaneous and the other according to which the present is what is experienced – a duration or temporal spread. He attributes this second image to the Phenomenological tradition, and he then quotes people like Husserl and Merleau-Ponty. In order to bring my interpretation of Husserl into focus I need to quibble with this distinction. The view that our experience of the present is durational – that it is non-punctal – is not something we owe the Phenomenological tradition, especially if we mean by this the movement begun by Husserl. Indeed, from Aristotle to Bergson and Whitehead philosophers of all sorts have been so taken by the non-punctal, non-instantaneous character of the experienced present that they have been

L. N. Oaklander (ed.), The Importance of Time, 171–177.

moved, in their metaphysics, to believe that the real present also is not instantaneous. The contrary position is taken by philosophers who think it is absurd to say a whole duration can exist at one time. They conclude our experience of the present must be specious in the sense that what we experience as present cannot really, as a whole, be present.

My quibble is that this issue predates, and is independent of, the Phenomenological tradition. People like William James and C. D. Broad, for example, tried to give theoretical accounts of the specious present. My more important point, however, is that we should be careful how we characterize Husserl's position. Let me be provocative: it is potentially misleading to characterize Husserl's position just by putting it in the Phenomenological tradition (even if he is a founder of that tradition). Why? Later folks in this tradition often take themselves to be discovering structures of experience that are independent of, or more basic than, traditional scientific pictures of reality. They sometimes view scientific pictures, especially mathematical ones, as mere abstractions or idealizations. Importantly, this is *not* true of Husserl's theory of time consciousness. He was a mathematician first, and it was for logical reasons that he was moved to construct his phenomenological theory in terms of instantaneous (non-durational) states of consciousness. In this regard, some later phenomenologists have more in common with Bergson than they do with Husserl.

II.

Professor Rubenstein correctly acknowledges this aspect of Husserl's theory when he characterizes Husserl's problem as the problem of accounting for how we can be perceptually aware of the immediate past, or perceptually anticipate the immediate future, in the strict present. As Husserl agrees, since the present of experience is specious and since neither the immediate past nor the immediate future exist at the moment of perception, it is mysterious how they are part of perception, or, indeed, part of any present consciousness.

Professor Rubenstein is also right to draw our attention to Husserl's diagram, illustrating the complex character of the momentary state of experience. (See his Figure 1, in the Appendix, p. 175.) At t3, the perceiver is in the instantaneous state CF, and whatever makes up CF is supposed to constitute the awareness of the temporally extended object (or process) AC. Here is where Rubenstein asks his critical question: what makes CF, or any sequence of CF's, a *temporal* representation? Since Kant, it has been widely agreed that a sequence of representations is not automatically a representation of a sequence. Husserl says that CF includes what he calls "retentions" (and it can include what he calls "protentions"), but Rubenstein wants to know what makes "the perceiver come to respond to what is in fact a temporally simultaneous array with a judgment about a temporal succession." He complains: "What is lacking, in short, is how such an array can give rise to such a temporally oriented judgment." (p. 164)

It is at this point that Rubenstein wants to go back to Kant for help. Rubenstein argues that temporal distinctions cannot be *derived* from the data in the array CF, so temporal distinctions must be *a priori*: "Time is required, but apparently cannot be derived from the unified representation given at t3... Time is required then as an *a priori* representation. And that of course is one of Kant's deepest points in the Transcendental Aesthetic" (p. 165).

III.

I want now to try to defend Husserl's theory from Rubenstein's complaint, and I want to use his complaint to show what I take to be the true character and value of Husserl's theory.

First, I think Husserl was very much aware of Kant's point that a sequence of representations is not a representation of a sequence. Husserl was very dissatisfied with psychologistic attempts to *derive* philosophical truths from the *contents* of experience, and very eager to posit *a priori* structures of experience when needed. So I think Husserl agreed with Kant that, in some sense, time is *a priori*. Moreover, I think he would agree with Kant that time is not just an *a priori* component of our judgments (that is, at the level of concepts and propositions, the categories need to be "schematized" in terms of time). More, at the level of pure receptivity time must be an *a priori* form of "inner sense."

But what does Kant's thesis mean? Like many transcendental theses it is wonderfully vague and vulnerable to idealistic hedges. How *exactly* is time a pure form of inner sense? Does the self-in-itself (supposedly nontemporal since time is transcendentally ideal) have some kind time "receptacle" waiting to line up sensations? Here, I think the virtue of Husserl's theory lies in seeing him as trying to go beyond Kant's insight.

How does Husserl do it? To make a long story short, he uses of the tool he got from Brentano – intentionality. In Husserl's program, the analysis of consciousness involves studying not just the objects of consciousness, but the character of the mental acts having those objects. Phenomenological reflection is supposed to make progress by revealing the necessary character of the acts (their distinctive kind of intentionality), rather than focusing just on their objects.

This is the key to answering Rubenstein's complaint. Husserl does *not* try to derive temporal *judgments* from the state CF. Rather, he argues that if CF is a conscious representation of a duration it must consist of a complex of intentional acts some of which are retentions and some protentions. And what makes these components retentions or protentions does not derive from their objects but is a matter of the distinctive character of their intentionality. That is, this intentional structure *is* the *a priori* form of experience. And this intentional structure is not just the form of linguistically expressible judgments – it is the elucidated form of Kant's inner sense operating at the level of sheer receptivity.

I have argued elsewhere that Husserl does not find this complex intentional structure by simply phenomenologically gazing at temporal consciousness. Rather, this structure is best understood as a theoretical posit (or as something transcendentally argued for) – it is what CF at t3 must be like if *it* is a conscious, perceptual representing of a duration. The figures in the Appendix are modifications and elaborations of Husserl's diagram which, I have argued elsewhere, better represents his intentions.

IV.

I will conclude by making a few comments on Professor Rubenstein's concern about how the future figures in temporal experience. It was one of Husserl's claims that much of consciousness is future directed, or anticipatory. He even made a part of *perceptual* consciousness future directed. But how can this be if the future does not exist to be perceived? Husserl is not a prophet; he's just a phenomenologist. We do not literally perceive the future. Rather, Husserl says, there must be some *component* of the complex perceptual act that is merely anticipatory, and he calls such components protentions. Again, what makes these components future-directed is their *a priori* intentional character, and again this can be viewed as an attempt to flesh-out Kant's form of inner sense in the idiom of intentionality.

It is interesting that there is a current controversy in cognitive science about whether real time perception can be computationally modeled or whether some "dynamic systems theory" will be needed. An example of the problem concerns the perception of a tennis ball by a tennis player: the player sees the ball as he moves to hit it. What does such seeing involve? What must it involve if the player is able to track the ball while moving and successfully hit the ball? His perceptual states must change very quickly and include some anticipation of where the ball will be. I think Husserl would be pleased if cognitive scientists someday posit perceptual modules or structures that play the role of his retentions and protentions.

APPENDIX

The following figures are from R. Hoy, "Time and the Mental: An Examination of Broad's and Husserl's Theories of Temporal Consciousness," Ph.D. dissertation, University of Pittsburgh 1973, University Microfilms 74-18, 421.

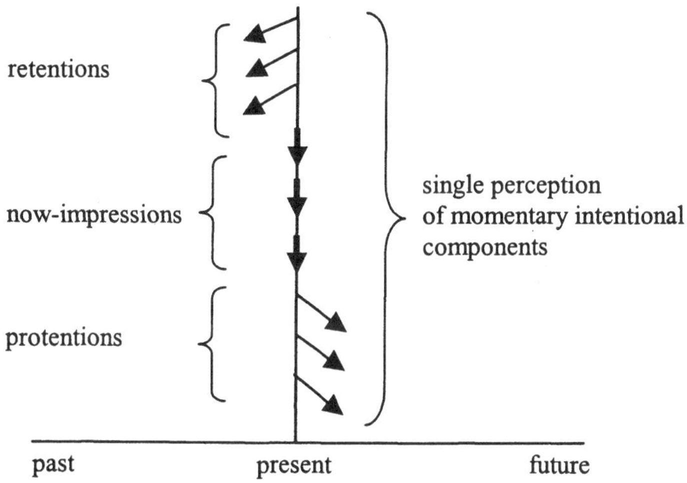

Figure 1

Each component of retentional consciousness is modified impressional consciousness. This account yields a more detailed picture of the double continuity of consciousness.

Here is the structure of the perceiving of a temporally extended object, a-b-c-d:

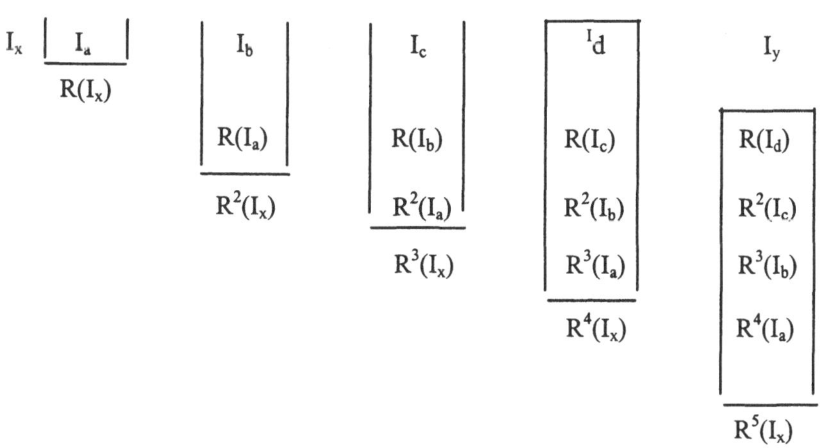

Figure 2

This picture is an elaboration of Husserl's rather obscure "diagram of time." It recaptures what Husserl's diagram shows, namely, the fact that the *same* temporal object, a-b-c-d, "sinks down" in consciousness. But this picture makes explicit, in a way that Husserl's does not, that the object sinks down in virtue of being given in different modes." For example, when the d part first appears, the a, b, and c parts are given in retentions of impressional parts:

I	→	d
R	→	I(c)
R^2	→	R(I(b))
R^3	→	R(R(I(a)))
R^4	→	R(R(R(I(x))))

Figure 3

It must be emphasized that one "sees" different aspects of this process depending upon whether one enters "transverse" or "longitudinal" intentionality. In transverse intentionality one is given the object, a-b-c-d, as temporally extended:

I	→	d				
R	→		c			
R^2	→			b		
R^3	→				a	
R^4	→					x

Figure 4

But constituting consciousness itself is given in longitudial intentionality:

I	→	
R	→	I( )
R^2	→	R(I( ))
R^3	→	R(R(I( )))
R^4	→	R(R(R(I))))

Figure 5

REFERENCES

Clark, A. (1988), "Time and Mind," *Journal of Philosophy* XCV, 7: 354-376.

Hoy, R. (1973), *Time and the Mental: an Examination of Broad's and Husserl's Theories of Temporal Consciousness, University Microfilms*, 74-18, 421.

Hoy, R. (1976a), "A Note on Gustav Bergmann's Treatment of Temporal Consciousness," *Philosophy of Science* 43, 4: 610-617.

Hoy, R. (1976b), "Science and Temporal Experience: a Critical Defense of C.D. Broad's Theory of Temporal Cognition," *Philosophy Research Archives* II.

Hoy, R. (1980), "Ambiguities in the Subjective Timing of Experiences Debate," *Philosophy of Science* 44, 2: 254-262.

Hoy, R. (1985), "The Given and the Self-Presenting," *Nous* XIX, 3: 347-364.

Husserl, E. (1966), in M. Heidegger (ed.), and J. Churchhill, (trans.), *Phenomenology of Internal Time-Consciousness* (Bloomington: Indiana University Press).

Part Four

GOD, TIME AND FOREKNOWLEDGE

15
ON THE ALLEGED METAPHYSICAL SUPERIORITY OF TIMELESSNESS

William Lane Craig
Talbot School of Theology

In his recent study *Time and Eternity*, Brian Leftow maintains that timeless existence is metaphysically superior to temporal existence (Leftow (1991), p. 278). He argues at great length that timeless beings have a higher degree of existence, in view of their truer presence and greater unity, than temporal beings and are therefore metaphysically superior to them.[1] It follows that it would most befit God, as a perfect being, to be timeless. In this paper I wish to assess Leftow's arguments in support of his claim.

1. THE MORE GENUINE PRESENCE OF TIMELESS BEINGS

The claim that timeless beings have more genuine presence than temporal beings is *prima facie* baffling since presence – or, better, presentness, to distinguish it from mere spatial presence, the opposite of absence – is the quintessential tensed temporal property. To ascribe presentness to a timeless being in any literal sense is patently self-contradictory, for if a timeless being had presentness, it would exist now, at the present time. As J. M. E. McTaggart, that patron of modern tensed theories of time, observed,

> The eternal is often spoken of ... as an "eternal present." As a metaphor this has ... some appropriateness, but it cannot, I think, be taken as more than a metaphor. "Present" is not like "existence," a predicate which can be applied in the same sense to the temporal and the timeless. On the contrary, its meaning seems to include a distinct reference to time .... (McTaggart (1909), p. 347)

Thus, I think that Eleonore Stump and Norman Kretzman are more accurate than Leftow in thinking of God's eternal present as God's specious present which covers all of time, a subjective rather than objective present.[2] When ancient philosophers spoke of the eternal in present-tense terms – an idiom

L. N. Oaklander (ed.), The Importance of Time, 181–186.

which passed into medieval theology, – this usage was probably in default of tenseless discourse, the inappropriateness of using past or future-tense language being evident. As Plato wrote, "we say of it [Eternal Being] that it *was* and *shall be*, but on a true reckoning we should only say *is*, reserving *was* and *shall be* for the process of change in time" (Plato, *Timaeus* 7, 38). Plato goes on to recognize the incongruity of using the present-tense "is" to speak of timeless eternity, but since no tenseless idiom was available, the present-tense was all that was left. This use of the present-tense with respect to the eternal led to metaphors like God's "eternal present" or even the *nunc stans* (static now) of eternity in contrast to the *nunc movens* (moving now) of time. If these metaphors are taken literally, they lead to absurdities. For example, God's eternal present must not elapse like the present instant, so it must somehow last, which leads to the malapropism of "atemporal duration." As Nelson Pike recognized, "if something *is*, but is such that it is never correct to say that it *was*, or that it *will be*, it is in a sense of 'is' that does not mean 'is now' " (Pike (1970), p. 15).

But Leftow defends the assertion that a timeless being is most truly present, whether one adopts a tensed theory of time or a tenseless theory of time. Tensed theories hold that attributions of pastness, presentness, and futurity to things/events express in some way an objective feature of reality, not merely the perspective or subjective impression of sentient beings. According to Leftow "on a tensed theory of time, to exist = to be present" (Leftow (1991), p. 84). Now while this assertion may characterize the view of some defenders of tensed theories, it is not true of all or perhaps even most. For many defenders of tensed theories would no doubt be quite prepared to admit, at least in principle, timeless existents into their ontology in the form of abstract objects, in addition to temporal existents. For such thinkers "to exist" may be tensed or tenseless, tensed in the case of temporal beings, tenseless in the case of atemporal entities. Only in the case of temporal beings is existence identical with being present. Thus, it is possible to be, but not to be present. Moreover, some defenders of tensed theories of time are quite prepared to say that past events exist and some even that future events exist, but that in neither case are such events present. Thus, "to exist" is tenseless and is not to be equated, in their view, with "to be present." Finally, those tense theorists who do equate existing with being present would not regard God, if He exists, as atemporal. Rather everything that exists, even abstract objects, exists temporally and is at some time present. Such thinkers would regard God, if He exists, as omnitemporal, as existing now, and as having a past, present, and future. None of these thinkers would agree with Leftow's claim that on a tensed theory of time

> if a timeless being exists, it *ipso facto* is in some sense present ... Hence a timeless being exists in a present to which nothing is past or future, and within its duration nothing is past or future. If this is so, then arguably, timeless beings are more genuinely present than temporal beings: for they are present without taint of past or future. (Leftow (1991), p. 84)

A present without past or future would be possible only if time were composed of a single instant, and then it would not have any duration at all. A being existing at such a present would not be timeless, but temporal and fleeting. To use such tensed words in any other way is non-literal or misconceived.

Leftow goes on to develop further non-literal senses of presentness such as character-presence and existential-presence, both of which have to do with immutability, not presentness. These notions do nothing to show that a timeless being is more genuinely present than any temporal being. Moreover, a metaphysically necessary, temporal God is as existentially immutable as a timeless being, and character immutability as defined by Leftow (having all at once all the attributes one ever has) does not strike me as a perfection at all.

Leftow then turns to a justification of God's being more genuinely present on tenseless theories of time. Tenseless theories are characterized by the denial that there are tensed facts about the objective world, that things/events really are past, present, or future. All events in the temporal series of events, whether past, present, or future from our perspective, are on an ontological par, and the presentness of things/events simultaneous with us is just a subjective feature of consciousness.

Leftow, however, fundamentally misconstrues the tenseless view of time. He thinks that

> On a tenseless theory of time, existence = presentness or pastness of futurity... The tenseless theorist will not grant that whatever exists at t is present at t. Tenseless theorists hold instead that whatever exists at t is present or past or future at t .... On a tenseless theory, to exist at t = to be present or past or future at t. (1991, p. 91)

This characterization is mistaken. The defender of tenseless theories of time does not believe that there are any monadic properties of pastness, presentness, or futurity. Hence, he does not equate existence with the disjunction of such pseudo-properties, lest there be no such thing as existence either. Such theorists hold that things/events exist without ever being past, present, or future. The disjunctive analysis of "to exist" is actually a technique of some philosophers of tense to parse the tenseless use of the verb in tensed terms: to say "x exists (tenselessly)" = "x has existed or x exists or x will exist." Leftow seems to have confused this device with tenseless claims.

Leftow is correct that tenseless theorists will not typically grant that whatever exists at *t* is present at *t*; but he errs in his positive characterization of tenseless claims, that such theorists hold that whatever exists at *t* is past, present, or future at *t*. At the most, some defenders of tenselessness hold that an event at *t* may be past for some sentient subject at $t'>t$, present for such a subject at *t*, and future for such a subject at $t^*<t$. Pastness, presentness, futurity, insofar as these are not reducible to the relations *earlier than*, *simultaneous with*, and *later than*, are subjective features of consciousness. Finally, to exist at *t* is not to be present or past or future at *t*, but just to be tenselessly located at *t*; tensed determinations have nothing to do with it.

These misunderstandings of the tenseless theory of time invalidate Leftow's further argument. He contends that because presentness has degrees, so does the disjunctive property of the three tense determinations. But philosophers of tenseless theories regard all of these properties as spurious. Similarly, Leftow proceeds to argue that the following claim is acceptable to such thinkers: "at t, existence located at t = presentness located at t." But such an identity statement is clearly false on tenseless theories, and the rest of the argument collapses. Hence, it seems to me that Leftow's attempt to justify the higher existence of a timeless being by means of its more genuine presence is maladroit.

2. THE GREATER UNITY OF TIMELESS BEINGS

What, then, of the purported greater unity of an atemporal being? Leftow claims that to exist is to exemplify a kind of unity and that since an atemporal being has greater unity than a temporal being, it has a greater degree of existence. Leaving aside the moot equation of existence and unity, we may ask why temporal beings have inherently less unity than timeless beings. Leftow's answer is that spatial objects are composed of parts and that "If the special theory of relativity is true, then whatever is located in time is located in space" (1991, p. 100; cf. p. 98).

But such an assertion is problematic for two reasons. In the first place, one could dispute the assertion on purely physical grounds alone in that it fails to take sufficient cognizance of the difference between coordinate time and parameter time. It is true that insofar as time plays the role of a coordinate, it is connected with a system of spatial coordinates, so that anything to which a temporal coordinate can be assigned is such that spatial coordinates are assignable to it as well. But insofar as time functions as a parameter, it is independent of space, and something which possessed temporal location and extension need not be held to exist in space as well as time. In Newtonian mechanics time plays the role of a parameter, not a coordinate, and, interestingly, the same is true of Einstein's formulation of the Special Theory Relativity – the now familiar space-time formulation derives later from Minkowski. The Special Theory can be validly formulated in either way. Moreover, since the Special Theory is a local theory only, we must, in order to achieve a global perspective, consider time as it functions in cosmological models based on the General Theory of Relativity, on which matter Leftow is silent. While time is defined in the standard Friedman models by means of spatial hypersurfaces, the time parameter in the Robertson-Walker line element which describes the space-time metric is distinguished precisely by its independence of space. Moreover, spatio-temporal coordinates in the General Theory are purely conventional and have no physical significance. Thus, it is not obvious that a being could not exist at certain moment of cosmic time without being spatially located as well.

But Leftow's argument suffers from a far more serious shortcoming than this. The argument appears to rest upon a crucial presupposition which will affect fundamentally one's theories of time and eternity and which I believe to be profoundly mistaken, namely, the reductionistic equation of time with *physical* time, that is to say, with time as it plays a role in physics. That this equation is mistaken is obvious from the simple fact that whereas physical time came into existence after the Big Bang singularity, time itself may well have existed prior to the initial cosmological singularity. A succession of mental events in God's mind – His counting, for example – would alone suffice to generate a temporal series in the absence of any physical objects whatsoever.[3] Thus, it is plainly not the case that something is in time if and only if it is in space – and that metaphysical truth is not negated by the fact that in physical theory an event which is assigned a temporal coordinate in space-time also has spatial coordinates as well.[4]

Leftow also asserts that a timeless being has greater unity than a temporal being because it "has its duration all at once," whereas a temporal being's duration "is dribbled out to it one instant at a time" (1991, p. 98). One could complain about these characterizations on the grounds that a timeless being has no literal duration at all and that temporal becoming need not proceed by instants. But these inaccuracies can be remedied without affecting the gist of the argument. Just substitute "existence" for "duration" and "serially" for "one instant at a time." Nonetheless, on an ontology of presentism, according to which the only temporal things/events that exist are those that are present, a temporal being which is present has all its existence which there is, and a timeless being can boast of no more unity than that. As for unity over time, this consists in its identity over time, its enduring as the same object through intrinsic change; this feat (which is impossible for a timeless being) is plausibly taken as testimony to its unity, rather than as detracting form it. On the other hand, if we adopt a tenseless theory of time, then it is false that the existence of temporal objects is dribbled out to them over time. Rather they just exist tenselessly at their (spatio-) temporal locations and their unity is just as great as that of a timeless being. Indeed, to borrow Leftow's terminology, relative to a timeless being the world-line of a temporal object (which many partisans of the tenseless theory take to be just the object itself) is extrinsically timeless, even though it is intrinsically temporal in that it has internally a structure ordered by temporal relations. Nothing comes to be or passes away; everything just exists (tenselessly). So it is difficult to see how a timeless being *qua* timeless would have any greater unity than a temporal being.

In conclusion, I do not see that Leftow has shown that timeless beings inherently possess a truer presentness nor a greater degree of unity than temporal beings. Hence, he has not demonstrated that a timeless God is metaphysically superior to a temporal God.[5]

NOTES

[1] Leftow (1991, ch. 5).

[2] Eleonore Stump and Norman Kretzmann (1992), p. 468.

[3] For good statements of the point see Grace M. Jantzen, (1984, p. 44); W. Norris Clarke, (1979, p. 94).

[4] The two contemporary thinkers who have signaled the distinction between time and physical time most clearly are Quentin Smith, (1993, Ch. 7), and Alan Padgett, (1992, Ch. 1).

[5] Leftow also claims that by examining the ways by which we distinguish dreams from reality, we discover several criteria of high reality which also serve to show that timeless beings have greater reality than temporal beings (1991, pp. 100-110). The procedure, however, seems to me to be misdirected, since the criteria are only criteria for our distinguishing reality from dreams; they are not properties which make the world more real than dreams. Leftow confuses the *ordo congnoscendi* with the *ordo essendi*.

REFERENCES

Clark, W. N. (1979), *The Philosophical Approach to God* (Winston-Salem, NC: Wake Forest University).

Jantzen, G. (1984), *God's World, God;s Body*, with a forward by John MacQuarrie (London: Darton, Longman, and Todd).

Leftow, B. (1991), *Time and Eternity* (Ithaca, NY: Cornell University Press).

McTaggart, J. E. (1909), "The Relation of Time and Eternity," *Mind* 18: 343-362.

Padgett, A. (1992), *God, Eternity and the Nature of Time* (New York: St. Martin's Press).

Pike, N. (1970), *God and Timelessness, Studies in Ethics and the Philosophy of Religion* (New York: Schocken Books).

Smith, Q. (1993), *Language and Time* (New York: Oxford University Press): ch. 7.

Stump, E. and Kretzmann, N. (1992), "Eternity, Awareness, and Action," *Faith and Philosophy* 9: 463-482.

16
PRESENTNESS, DATES AND ETERNITY

Brian Leftow
Fordham University

Craig takes up an argument from my treatment of some claims of Augustine. Cleaned up a bit, it runs thus:

(1) existence = presentness.
(2) a timeless being exists.

So,

(3) a timeless being is present.
(4) no part of a timeless being's life has, had or will have a past or a future.
(5) any being no part of whose life has, had or will have a past or a future is more fully present than any being some part of whose life has, had or will have a past or future.
(6) Some part of any temporal being's life has, had or will have a past or a future.

So,

(7) any timeless being is more fully present than any temporal being.

So,

(8) any timeless being exists more fully than any temporal being does (from 1, 7).

The premises, (1), (2) and (4)-(6), are claims Augustine makes. If they make sense, they make sense of (8), and so of Augustine's talk about degrees of existence. If they are plausible, so is (8). (4) is true by definition and Craig accepts (6) elsewhere, so I won't discuss these.[1]

L. N. Oaklander (ed.), *The Importance of Time*, 187–197.

PRESENTNESS AND TIMELESSNESS

Craig claims that (3) is "self-contradictory," because "if a timeless being had presentness, it would exist now, at the present time." But it's unclear why we are supposed to believe Craig's conditional. Some might suggest: because it's an instance of

(9) if something had presentness, it would exist now.

But (9) appeals to us, if it does, as some sort of conceptual necessary truth. And it is not one: it may be false. Some things may be present without existing now. There may be time-series discrete from ours, i.e. series none of whose moments occur before, after or while any moment in our own time-series occurs. If this *is* possible, as I think, then there may be a present in another time-series. Yet what is present in that time-series does not exist *now*, i.e. at the present of our series. Times in a time-series discrete from ours have no temporal relation to our times at all. Again, many take seriously the thought that time could "branch," i.e. that what had been one time-stream C might break into two streams AB with the unsplit part of C as a common past, such that no instant of A after the last common past instant occurs before, during or after any instant of B, and *vice-versa*. If time branched, there would after the branch be two presents. At either, (9) would be either false or irrelevant, depending on how one read its tenses. Read as

(10) if something tenselessly-had presentness at some present, it would exist now,

it would be false, for items in the other branch would have presentness but not exist *now*, at *this* branch's present.[2] Read as

(11) if something presently-had presentness, it would exist now,

it would be trivially true – and not count at all against the claim that other things tenselessly-might have presentness at another present. So the claim that something has a present which is not in our past or future but is not *our* present is no self-contradiction unless the claims that there is a second time-series or a branch in a time series are self-contradictions.

Perhaps we're supposed to believe Craig's conditional because it follows from

(12) any present belongs to some time-series

plus the presumably-true thesis that ours is the only time-series. But Craig gives no argument for (12), unless perhaps in saying that

> A present without past or future would be possible only if time were composed of a single instant, and then it would not have any duration at all. A being existing at such a present would not be timeless, but temporal and fleeting. (p. 183)

This puzzles me. If a being ceases to exist, first it exists and then it does not. While it existed, its non-existence was in its future. If something has no future, it has no future non-existence to become present. So it cannot fleet; something in a present with no future cannot cease to exist. And why would one instant, not earlier or later than anything, be temporal rather than something else? Craig is entitled to call an isolated present instant an instant of *time* only if he has already shown that any present must belong to some time-series. This passage is the closest he comes to doing so, and if it tries, it begs the question.

Perhaps Craig thinks (12) too obvious to need arguing. Well, a lot of fairly smart folks have not found it obvious at all. It behooves us to ask why.

Pace Craig, this began with Plato. At the place Craig cites, far from calling it "incongruous" to use the present tense of eternity, Plato writes of the timeless that not "was" or "will be" but "'is' alone is *properly* attributed to it" (Plato, *Timaeus*, 38a; my emphasis), and that the present tense is improper *when applied to temporal things*: "when we say that what has become is become and that what becomes *is* becoming. . . all these are inaccurate modes of expression" (*Timaeus*, 38a). If Plato meant to call it improper or inaccurate to use the present tense of the timeless, he'd be claiming that the present tense could never be used properly and accurately at all. But rather than discuss Plato's reasons to say that a timeless being has a present, I'll take up the Bible and Boethius.

Biblical authors seem to treat the claim "God is eternal" as very roughly equivalent to "God always exists."[3] "God always exists" answers the question "When does God exist?" As an answer to a when-question, "God always exists" uses a predicate which belongs in the Aristotelian category of "when." So then does the Biblical claim "God is eternal."

Boethius' *de Trinitate* investigates how terms in the Aristotelian categories apply to God. When he turns to the category of "when," Boethius equates "God always exists" and "God is eternal," takes "God always exists" as his sample when-statement, and explicates it in terms of God's timelessness.[4] So for Boethius, even if God is timeless, "God is eternal" answers the question "when does God exist?" It does not reject this question. If he thought it rejected the question, Boethius would say that no terms from the category "when" apply to God.

A date is an answer to "when?" Thus for Boethius, even if God is timeless, "God is eternal" gives a date for God's existence. For Boethius, eternity is a date, logically speaking. Given Boethius' position as a source for later expositions of God's eternity, this point applies throughout the Middle Ages. At least two things follow from Boethius' claim that eternity is a date.

First, if eternity is a date, God has a present. On a tenseless theory of time, if I am alive on Dec. 28, that of my life which is at Dec. 28 is present relative to Dec. 28. If time is tensed, then on Dec. 28, Dec. 28 and what is at it are present in some further, non-relative sense. So for a tenser, if one is alive at a date, that of one's life which is at that date is *simpliciter* present when it *is* that date: if I am alive on Dec. 28, that of my life which is at Dec. 28 is *simpliciter* present when the date *is* Dec. 28. Again, if someone's life is dated, the following seems literally true: the present of that person's life at a given date is that of it which is actual at that date and not past or future at that date. If God lives at a date, that of His life which is not past or future at that date is present at that date.

Secondly, if God has a present, the right way to talk about God's life is tensed, not tenseless. For we use tenses to express relations between our present and other presents. "I will eat that sandwich" makes *inter alia* the claim that the time at which my eating the sandwich will be present – the present of my eating – is in the future. If a timeless God has a present (as my eating the sandwich will), the right use of tense can indicate the relation between that present and ours. Tensed talk would be no metaphor here. It would do for talk of a timeless God what it does elsewhere. As a gesture toward the unusual tense needed, we might attach "then" without antecedent to a present-tensed verb. As "Putnam is writing" places Putnam's writing in our present, "God isthen thinking" would say that at God's own present, God thinks. "Isthen" would imply no relations to our time: the truth of "God isthen thinking" would not entail that He is thinking now, or at any other of our times.

These points turn on the claim that eternity is a date. So if we want to evaluate the claim that a timeless being has a present, we must examine the date-claim. I now argue for it.

AN ARGUMENT FOR THE DATE-CLAIM

It seems to me that a term x functions literally as a date-term if a sentence is true which has the form "because p tenselessly-is true at x, at t, p was already true." Now consider the claim that

(13) it was true from all eternity that 2+2=4.

(13) is true. (13)'s force is that at any time, it was already true that 2+2=4. (Lucas captures the thought that logical truths are true from all eternity by the convention of dating them to a date $-\infty$[5]) So (13) implies that

(14) because it tenselessly-is true from all eternity that 2+2=4, (t)(at t, it was already true that 2+2=4).

If (13) is true and (13) implies (14), (14) is true, and so "from all eternity" is a date-term. We have no qualms granting instantiations of (14) for any time after time's first instant, if there was one. But consider even a supposed first instant. We would not want to say that it *became* true at that instant that 2+2=4. For this would have been true even if time never existed. If so, it seems that we should say that at that instant, it was already true that 2+2=4. If we should say this at the first instant, then of course we should say it at every later instant.

The first-instant "already true" has roughly the force of "true, and would have been true even if time had not existed." At each later instant, a further note is added, that it *was* true earlier. But while "already true" does not entail at the first instant all that it entails at later instants, its meaning may well be the same. For all t, a proposition is already true at t just in case it is true at t and would have been true had time never reached t. At the first instant, being already true implies being true even if time had not existed. At later instants, it does not imply this. At any instant after the first, being already true implies that the proposition was true at an earlier time. At the first instant, it does not.

"From all eternity," then, functions literally as a date-term in (13) and (14). Now had time not begun, it would have been the case that 2+2=4, but not at any time – and so timelessly. That 2+2=4, then, is a timeless truth at least if time does not begin: and "from all eternity" allows for this. But suppose that time had not begun. Then that 2+2=4 would have been timelessly true, but in one respect *not* true "from all eternity." For there would have been no later time from which (as it were) to look back to eternity, as "from all eternity" suggests. Instead, it would simply have been true *at eternity* that 2+2=4. Eternity would have been the date of this truth, logically speaking: "at eternity" inherits the dating function of "from all eternity" if the dates involved collapse into sheer timelessness. So if time had not begun, eternity would have been the date of all timeless truths. But the contingent fact that time exists cannot make a term change its logical category. If there is a possible circumstance in which "at eternity" is a date-term, it is a date-term *simpliciter*. So logically speaking, eternity is a date. *Per* earlier argument, this gives us a case for (3) which is independent of (1). This is worth having, as Craig is correct that tensers who deny Augustine's presentism – (1) will reject the Augustinian case I made for (3).

Though I've spoken as if we looked back to eternity, none of this places eternity in some fictive infinite past, as Lucas imagines it.[6] A better analogue is the present of a discrete temporal system. Say that at some time in some other time series,

(R) Ruler is President of Ruritania.

What should we say about (R)'s status as our time begins? Not that it *is* true at that time, or that it will become true. Nor is (R)'s truth then in our time's past – at its first instant, our time *has* no past. If (R)'s series is discrete from ours, (R)'s truth has no temporal relation to our time. Rather, at our series' first instant, (R)'s truth is a given, but without any temporal relations to our time, and would have been given if our time had never begun. This is just what it means to call a timeless truth "already true" at time's first moment. But again, in that other time series, (R) expresses (at some time) a *present* fact.

FROM TENSED TO TENSELESS VERSION

Augustine held (1), and so a tensed theory of time. I have argued (though not quite in these words) that one can de-tense the Augustinian argument by *inter alia* substituting for (1) and (3)

(1a) existence = being present, past or future, and
(3a) a timeless being is present or past or future.

(1a) is a tenseless account of existence because with existence so understood, "A exists" licenses no inferences as to when A exists, or which of past, present or future A's existence is. I used (1a) to transfer Augustine's reasoning to a tenseless context, and claimed that the resulting argument preserves for de-tensers most of whatever force the tensed version has.

Craig asserts that I confuse (1a)'s disjunctive account of tenseless existence with some other thing – perhaps the tenseless "is" which tense logic places within the scope of sentential tense operators. This is provably false; I have an article in print which hinges on the difference between the two (Leftow, 1991). I just *like* the disjunctive account better. Most or all verbs are tensed in ordinary usage.[7] (Craig may concede as much in asserting that Plato had no tenseless idiom available.) So (1a) explains the meaning of the tenseless "exists" in terms close to ordinary usage.

Craig also asserts that no de-tenser would accept (1a). True, de-tensers don't believe in irreducible monadic unindexed tense-properties, or give (1a) as their account of existence when in church with fellow believers. But this hardly "invalidates" my further argument, or even entails that de-tensers would reject (1a). De-tensers understand tenses. They just reinterpret them, and not necessarily with reference to minds, either. A de-tenser will take e.g. "is present," unindexed, as shorthand for "is present at some time t." If the de-tenser goes no further, then for all Craig says, the de-tenser can accept (1a) in like guise and also accept any reasons I can give for taking presentness as degreed, and so the rest of my *T&E* argument. De-tensers can however go further, reduce tense to location, and say e.g. that for any event E,

E is past at t =df. E is located earlier than t, and
E is future at t =df. E is located later than t.

To a de-tenser, of course, being earlier than t is as legitimate a property as being south of Boston. If this is how a de-tenser would treat pastness and futurity, then the parallel tenseless account of presentness should be that

E is present at t =df. E is located at t,

Craig writes that on a tenseless theory "to exist at t is... just to be tenselessly located at t." But surely on a tenseless view, anything anywhere in time exists, and so at t exists, i.e. exists from the standpoint of t. On a tenseless view, items can exist at t without being located at t: on a tenseless view, today, Ulysses S. Grant exists, in the past which (on a tenseless view) is as real as today is. To be located at t is the tenseless equivalent of being *present* at t.

If this is how a de-tenser would read an argument including (1a) and (3a), the real question – one Craig doesn't pose – is whether Augustine's argument preserves whatever force it has given this further treatment of its premises.

To get a de-tensed (1)-(8) which a de-tenser would not blush to say in church, we can substitute for (1a)

(1b) A exists =df. A is located earlier than, later than or at some date d,

substitute for (3a)

(3b) a timeless being is located earlier than, later than or at a date d,

and substitute for (4)

(4a) there is no date than which any part of a timeless being's existence is earlier or later.

(4a) is unproblematic on any account. De-tensers would have with (3b) at most problems about eternity's being a date, paralleling those tensers might have with calling a timeless being present. Given Craig's survey of the variety of tensed views, the proportion of de-tensers accepting (1b) might exceed that of tensers accepting (1). So these steps would have for de-tensers at least the plausibility their tensed analogues have for tensers.

Craig never discusses (5), but only whether a de-tensed version of (1)-(8) is possible. Now *T&E* argues with Augustine that presentness is degreed, and that if it is, then so is a disjunctive property including presentness. The latter point remains true whatever we make of presentness. We may wonder, then, whether the reduction of presentness to location leaves any reason to see presentness as degreed. Are there degrees of location? Can one thing be more fully located in a place or time than another? Here are some common opinions: a student who pays attention in class is more fully there than one sleeps through a class. The latter, we sometimes say, is hardly there at all. One who lives "in the moment" is more fully there, in the moment, than someone always regretful about the past or worried about the future. (Craig and I are theists, and both [I take it] aware of a long tradition of making sense of God's presence in the world by talking *inter alia* of His awareness of it. If awareness is irrelevant to being located, then the doctrine of God's omnipresence is that much the closer to being an idle metaphor: a consequence I assume Craig would not welcome.) Someone whose family has roots in a town, who has lived there all his life, is more fully there than someone just passing through. The University is more fully in the city if all the University buildings are there than if some are there and some are in the surrounding suburbs. Obviously, all these claims deserve a good deal of analysis. My point in bringing them forward is simply that there may be more than one dimension to location, and more intuitions to take account of than just the flat-footed one which says "either the location encloses the thing or it does not: there's no question of degree." The case for degrees of location remains open. And this leads me to the one remark I'll make about de-tensing (5): if being present without taint of past or future, having a life which is purely and only present, really does entail being more fully present, then in the same way, being located at t without having a location before or after t may entail being more fully located at t. One intuition behind the first claim is that not also having parts elsewhere makes an item more fully here. That intuition can motivate the tenseless version too.

Turning now to what Craig has to say about the greater unity of the timeless, Craig notes that I take time as it figures in physics – "physical time" – to be, simply, real time, and writes

> That this equation is mistaken is obvious from the simple fact that whereas physical time came into existence after the Big Bang singularity, time itself may well have existed prior to the ... singularity. A succession of mental

> events in God's mind – His counting, for example – would alone suffice to generate a temporal series in the absence of any physical objects whatsoever. Thus it is plainly not the case that something is in time if and only if it is in space. (p. 185)

If God existed and was temporal before the Big Bang, then perhaps time existed then but physical time did not. (Perhaps not. Couldn't the time God existed in be "time as it plays a role in physics," i.e. the same time which was to figure in physical theory once there was a physical universe? We can at least *picture* a line, a one dimensional figure, by itself for part of its length, and for part of its length one axis of a 4-space.) But that this *would* be true *if* God was as described hardly entails that it is not the case that something is in time if and only if it is in space. Before drawing this conclusion, there is the little matter of showing that God *was* in this position to deal with. In fact, Craig does not think He was. Nor do I. So why was this worth noting? Craig's raising the *possibility* that God have a sequence of mental events before Creation suggests that perhaps he means to argue this: it's *possible* that God be in time before Creation, so it's possible that something be in time but not space. But I do not deny that something could be in time but not space: my claim in *T&E* was only that *in fact*, whatever is in time is in space. So the modal version of the argument fails to connect.

Whether existence be tensed or tenseless, a timeless God has His life all at once: all parts of His life are identical. Something which exists through some continuous stretch of time has only a proper part of its existence during any proper part of its time. It is irrelevant whether becoming proceeds by instants. There is during any proper part of a continuant's time only that of its existence which is just then, and this is not *all* the existence it has had or will have. On a tenseless view of time, the "dribbling" of existence is just the same: a temporal existence's having distinct slices or proper parts, and only one slice or proper part's being in each proper part of its time. It is an existence's being in many locations, and in all but one (its entire time-span) being there by having a proper part or slice there rather than by a 1:1 relation between the whole duration and the location. Because the difference with respect to dribbling is a difference in way of location, it is one with respect to way of existing, as tenseless theories conceive this.

One might reply: so what? How else could things exist save part here, part there? Well, a thing could have its whole existence in any proper part of its site, as timeless or instantaneous beings do. An instantaneous temporal being does have its duration all at once, thus never missing any of its existence. But it buys this property at great cost, that of missing all but an instant of time. Missing most of time certainly seems to be an imperfection: we would rather not do it. Augustine and Boethius claim that a timeless being has the best of both worlds – never failing to possess any part of its life, yet never ceasing to exist and somehow having all of time a "given" for it, as all of a second time-series' truths are "givens" for us, within our time-series. If it is an imperfection to at some points in one's life lack what is at other points in one's life, or to miss some of time, or to cease to exist, then even if time is tenseless, the difference between being temporal and being timeless is relevant to a being's perfection. Craig asserts us that for presentists, "a temporal being which is present has all its existence which there is." True, but the non-presentist tensers of whom Craig reminds us could not accept this claim, at least with "is" tenseless, to range over the past and/or future which (Craig reminds us) they think to also exist. With "is" tensed, they'd have to call the claim irrelevant: sure, it has all the existence there *presently* is, but on their terms, that's not all there is *simpliciter*. And even a presentist must grant that an enduring temporal thing at no time has all of its existence which there was or will be. A timeless being suffers no such deprivation.

Now given only what I have said so far, a higher degree of presentness is *part* of what makes timeless existence more perfect, but on its own is equivocal- for given only what I've said here, instantaneous events (if any) are more fully present than continuing events or continuants are, and yet we would not call instantaneous events' way of being more perfect than these others'. But all this just involves *one* gloss on higher-degree presentness. *T&E* treats more than one kind of presentness, and treats highest presentness as the conjunction of these kinds' intrinsic maxima. A full recasting of the other kinds, worked into the present account, would (I think) warrant the conclusions that highest-degree presentness is a perfection on its own and that instantaneous temporal events do not in fact have it, nor even an overall degree of presentness greater than continuants'.[8]

NOTES

[1] William Craig (1994), pp. 29-30.

[2] I indicate below a preference for taking tenseless claims as disjunctions of tenses. One can do this with the sort of tenseless claim I've just made, if one introduces tenses suited to talking of times in the other branch. I think this latter quite doable.

[3] For a careful treatment of the Biblical witness, see my "Eternity and Immutability," (forthcoming).in Willliam Mann (ed.), *The Blackwell Companion to Philosophy of Religion*.

[4] Boethius, *De Trinitate*, in Stewart and Rand, *Boethius: The Theological Tractates* (1926), p. 20.

[5] J. R. Lucas, (1989, p. 114).

[6] Though it does sit well with Hasker's and Zagzebski's notion that eternal facts have the "hardness" of past facts.

[7] Note, incidentally, that if eternity is a date, and is the date of timeless truths, then even verbs involved in stating timeless truths are tensed.

[8] Some other points are worth noting in passing. First, it's surprising to hear McTaggart called a patron of tensed theories of time (p. 1). True, McTaggart held that if time exists, it is tensed. But he is better known for holding that any tensed theory is incoherent, and concluding that there is no time. With patrons like that, one doesn't need enemies. Again, it's almost as surprising to hear Stump and Kretzmann said to treat God's eternal present as specious and merely subjective (1-2). In the article Craig cites, they introduce the specious present as an *analogy* to support the claim that God's eternal present is objectively both extended over the whole of time and indivisible. They nowhere give this claim up; if they meant to speak only of a subjective, specious present, there would have been no need to go on to define an *objective* relation of simultaneity between God's present and others ("ET-simultaneity"). Finally, the claim that I confuse *essendi* and *cognoscendi* (n. 15) is incorrect. *T & E* explicitly distinguishes marks of highest existence/unity from features *constituting* highest existence/unity (99), argues explicitly for a connection between the marks and the underlying features (99-100), throughout distinguishes *criteria* (a plainly epistemic matter) from either (100-110), and finally argues on 110 that the criteria, developed independently, *are* appropriate for placing items on an objective scale (since they do in fact function appropriately in determining the *objective* non-existence of what we wind up calling dream objects). The logic of the placing argument is this (see 99-100 for a parallel): I tolerate just a three place scale: non-existence, inferior reality, superior reality. So any criterion which is reason to treat an existing object as more or less real than another suffices to determine its place on the scale. So if the dream/reality criteria do also apply between temporal and timeless, they suffice for placement (see 101-102). There is no confusion here, but just a way to use criteria evident to us to make a judgment about the *ordo essendi*.

REFERENCES

Craig, W. (1994), "The Finitude of the Past and the Existence of God," in William Craig and Quentin Smith *Theism, Atheism and Big Bang Cosmology* (New York: Oxford University Press): 29-30.

Leftow, B. (1991), "Timelessness and Foreknowledge," *Philosophical Studies* 63: 309-325.

Leftow, B. (forthcoming), "Eternity and Immutability," in Willliam Mann (ed.), *The Blackwell Companion to Philosophy of Religion* (Oxford: Basil Blackwell)

Lucas, J. R. (1989), *The Future* (Oxford: Basil Blackwell).

17
PARTS, WHOLES AND ETERNITY

Brian Leftow
Fordham University

For Western theists, God is eternal. Western theists disagree over what it is to *be* eternal, but the most influential view historically has been that of Boethius, who wrote that "eternality is the complete possession, all at once, of illimitable life ... for which nothing of the future is absent and nothing of the past has flowed away."[1] On one reading, Boethius' account entails a surprising view of how God's life fills time. I now set this view out, and take a step toward showing that it is at least coherent.

BOETHIAN ETERNITY

My life fills a region of time – the past or future – by having a part in that region. If for God "nothing of the future is absent," none of God's life is yet to come: God's life has no future part. If for God "nothing of the past has flowed away," none of God's life is ever over: God's life has no past part. If these things are so and God presently exists, that of His life which is present is not a proper part of it either. For Boethius, then, God's life has no proper temporal parts. On one reading, Boethius holds that though it has no temporal parts, God's life nonetheless *lasts* or *endures*, i.e. is extended as lives in time are.[2] Exegetes differ on whether Boethius really does assert this. I do not want to argue exegesis here. I want instead to suggest that we can make some sense of a life's filling time yet having no temporal parts.

Let us take it that for Boethius, God's life fills stretches of time but lacks proper temporal parts: though it fills the past, it has no past part, and though it exists at earlier times, it has no earlier parts.[3] A Boethian eternal life, so taken, is an extended temporal simple. The claim that there could be such a life is counter-intuitive at least three ways. For one thing, many think it just obvious that being extended implies having parts.

Again, on Boethius' account, the time God's life fills divides into parts (days, years), with boundaries between them, and yet God's life has no internal boundaries or parts. So Boethius denies that filling or covering a region with internal boundaries and parts entails having internal boundaries

L. N. Oaklander (ed.), The Importance of Time, 199–206.

and parts. He claims, in other words, that the topology and mereology of a region need not determine the topology or mereology of objects filling the region. But if the space an object occupies has a left and a right half, for instance, it is hard to see how the object in the space could fail to have left and right halves.

Finally, the claim that something is extended but has no proper parts can seem to have weird consequences. Zimmerman sketches the oddity of a spatial parallel:

> Consider an extended simple that is supposed to fill an extended region. If it failed to have a part filling each sub-region, then each sub-region would be empty – and how could the whole region be filled, if it were just a sum of empty regions? If ... no sub-region of the region occupied by an object could fail to be filled ... then the simple must be present "the whole in the whole and the whole in each part" .... But then no attempt to interact differentially with an object filling just a proper part of the region could be successful. When one sees the top half of the thing, one thereby sees the bottom half as well; when one touches the left side, one touches the right also. Perhaps this bizarre mode of space-filling is not absolutely impossible; but surely the more sensible kinds of extended objects ought not participate in it. (Zimmerman (1996), p. 154)

There thus is ample reason to doubt that Boethius sketches a real possibility, and say that no object can be both extended and simple. Yet there is also reason to hope that there *can* be extended simples, at least if one would like it to be possible that there are now ultimate particles of matter with positive mass. No real physical particle outside a spacetime singularity could be a point-mass. For any positive mass, in an object of zero extent, would be infinitely dense. So if no particle outside a singularity can be infinitely dense, any ultimate particle with positive mass and outside a singularity would have some positive extent. Yet an *ultimate* particle is by definition a literal atom: it consists of no more basic particles, and physically cannot be split. So any particle of this kind would be an extended physical simple. It would relate to the space it fills as Boethius thinks an eternal life relates to the time it fills: it would not have physical parts corresponding to the parts of the space it fills. As a Boethian eternal life fills past time without having a past part, our ultimate particle would fill (say) the left half of its space, but there would be no physically separable object which just fills the left half of its space. One wonders, then, whether our particle would nonetheless have spatial parts. Would it have a left half, even though it physically could not be split in half? Would it have them if this physical necessity were also a metaphysical necessity, one grounded in its nature?

Some might reply that even if the object cannot be divided, the region it occupies can, and this entails that the object has spatial halves. But surely an object's part-structure is an intrinsic matter, something settled purely by the object itself. If so, why should the fact that the object occupies a parted space entail that the object has spatial parts? The space an object occupies, after all, is not clearly an intrinsic feature of the object; to occupy a space is to be related to something else. And to treat space as having a part-structure it can "force" on its occupants bids fair to commit one to a strong substantivalism about space. If space's part-structure is instead dependent on what stands in spatial relations, then a given space's having parts may not entail that what now occupies it has parts, for the space's having parts may depend not on this, but on the fact that distinct objects might each overlap parts of the space. If we are not antecedently committed to substantival space, then, perhaps we ought to consider whether there might be extended, wholly simple ultimate particles. Thus Boethius raises a question which should interest not just theists, but also believers in ultimate particles.

PARTS AND FILLING

Let us say that an item A has a proper part B just in case there exists an item B, distinct from A, such that all parts of B are parts of A, but some parts of A are not parts of B. To approach Boethius' view, let us distinguish two ways items can fill regions.

Consider a cueball filling a region of space. The ball fills the region if it has a part filling each part of the region – a left half in the region's left half, a right in the right, etc. If this is how the ball fills the region, it fills the whole region and its parts fill the region's parts. So let us say that such an object fills its region by parts. If a thing fills a region by parts, the region's mereology is also the thing's. There is a bounded part of the thing for each bounded part of the region.

This does not seem to be the only way a thing could fill a region. We could say instead that the ball fills its region *and its region's parts*, but fills the parts *itself*, rather than by having a part of itself in each part of the region. That is, we could say that only the ball itself, not any part, ever bears the "fills" relation to the ball's region or sub-regions. Let us say that such a ball fills its region primarily. If the ball fills its region primarily, the region's mereology is *not* the ball's. The region has bounded parts, but there are no corresponding boundaries or parts within the ball. I now explain filling-primarily.

It may seem that if our ball fills a sub-region of its region, it *must* have a distinct part in that sub-region, and the ball itself *cannot* fill any such sub-region. But those who think this tacitly take "the ball fills the sub-region" as "the ball *just* fills the sub-region," i.e. as implying that the ball fills that sub-region and fills no region that sub-region does not wholly include, and that the sub-region's boundaries bound what fills it. But a ball can fill a region it does not *just* fill. For a ball can *overflow* that region. The Great Plains fill Kansas. But the Great Plains do not *just* fill Kansas. They overflow Kansas, also filling Nebraska and other states. There is a primitive, dyadic relation of filling. It is a determinable with at least two determinates, *just filling* and *overflowing*. We can define overflow this way:

(1) A overflows region B =df. A fills B and A fills some region which is not part of B.

We can also define just-filling in terms of filling and parthood:

(2) A just fills region B =df. A fills B, and A fills no region which is not part of B.

So overflowing and just-filling are at least sibling concepts. One is no more basic than the other. In fact, given (1) and (2), they are interdefinable:

(3) A overflows region B =df. A fills B and A does not just fill B.
(4) A just fills region B =df. A fills B and A does not overflow B.

Given (1)-(4), one cannot claim that taking overflow rather than just-filling as the basis of an account of how some object fill some regions gets some sort of conceptual priority wrong.

One can define filling by parts this way:

(5) A fills region B by parts =df. A just fills B, and for every proper sub-region C of B, some proper part of A just fills C.

(5) appeals only to just filling, and requires A to have proper parts if its region does. If an item fills its region primarily, the account goes this way:

(6) A fills region B primarily = df. A fills B, and for every proper sub-region C of B, A overflows C.

(6) appeals only to overflowing and a notion which occurs in overflowing's definition. It does not require A to have proper parts.

(6) seems as clear and as conceptually parsimonious as (5). Nothing in the two definitions seems reason to prefer a metaphysic of filling by parts to one of filling primarily, and parsimony favors the latter in that it avoids profligate commitment to parts. The mere availability and clarity of (6) are

enough (I submit) to block any facile inference from "A's region has a left half" to "A has a left half." To make someone accept that inference, one must give some reason to believe that A fills its region by parts rather than primarily. What is more, nothing seems to rule out *a priori* a *mixed* metaphysic of filling: perhaps there could be an item with parts corresponding to only some of its sub-regions. Why not? Absent some showing that both a pure metaphysic of filling-primarily *and* a mixed metaphysic of filling are untenable, the only way to make someone accept the move from "A's region has a left half" to "A has a left half" is to give some independent reason to believe that A has a left half. If so, the fact that A's region has a left half has no independent force at all toward settling whether A does.

SOME PUZZLES

To see whether a metaphysic of filling-primarily could make sense, let us take up Zimmerman's puzzles about extended simples for a partless cueball, O.

O, as it is partless, does not have a distinct part filling any part of its region. But it does not follow that its region's sub-regions are empty. Rather, O itself fills and overflows each sub-region. Nor does this imply that each sub-region somehow wholly contains O. True, only O does the filling of each sub-region. But a region wholly contains O only if O does not overflow that region.

If O overflows a sub-region, this does not consist in O's having parts in other sub-regions. Rather, O itself overflows each sub-region because O itself also overflows other sub-regions. Conversely, that a sub-region partly contains O does not on our account entail that the sub-region wholly contains a part of O. If O is partless, then if a sub-region partly contains O (i.e. if O overflows the region), there is nothing the region wholly contains – and yet the region is not empty.

A partless thing would not *have* a top or a bottom half, or a left and right side, if those phrases name parts. So it would not be true that if one sees O's top, one sees its bottom, or if one touches O's left, one touches its right. On the other hand, a partless thing would occupy a *region* with top, bottom, left and right parts. One could certainly see O overflow the top sub-region it occupies (namely, where it occupies some part of its bottom sub-region) without seeing it overflow the bottom sub-region it occupies (namely, where it occupies some part of its top sub-region), or touch O where it overflows the right sub-region it occupies without touching it where it overflows the left sub-region it occupies. If O fills its region primarily, "O's right half" has the sense "O where it fills the right half of its region." So even if O has no parts, it is not true that by touching O's right half, one touches O's left.

If this is so, then in a sense, we never touch just part of O, but instead always touch O, whether we reach to O's right or O's left. But this is not particularly odd. If I see an apple, we may say that I see (say) the apple's front half, and by seeing that see the whole apple. But we need not. For I do not see the apple's front half *as* the front half. I see it as an apple where it faces me, i.e. as a whole apple insofar as it is in a given region.

Our ordinary ways of speaking reflect this. If I say "I see part of an apple," most ordinary speakers will take me to mean something like "I see a slice of an apple" or "I see an apple peeking out from a pile of lettuce." This is because we usually take "I see an apple" to entail that I see a whole apple; as this is so, saying that I see a part usually suggests that there is less than a whole apple there or that there is some obstruction, so that less of the apple than usual is directly presented. True, the whole of the apple is never directly presented to me: that is, I always see the whole *by* seeing the apple occupy just part of its region, and never see every part of the region the apple occupies or the apple occupying every part of its region. But this does not entail that I do not see the whole apple. So it is not clear that properly describing my seeing the apple even *prima facie* commits us to there being an apple-part I see. In fact, the argument that since I do not directly see the entire apple, there must be an apple-part which I do directly see, is more than a bit like faulty arguments for the existence of sense-data.

Still, a worry remains. It may be acceptable to say that I always see the whole apple, not an apple-part. But I do not always see all of the whole apple. I am not directly presented with its interior. Suppose that I touch O. Surely I touch O's surface and not O's interior. If so, O has a surface and an interior – and so has parts. Again, surely (one wants to say) my finger does not touch O's whole surface. So surely there is that of O which my finger touches and that of O which it does not. If so, O has parts. If not, doesn't it follow that the whole of O somehow squeezes into the surface-area my fingers touch – and so too into the place I touch next, and the one after that?

I reply this way. My finger clearly can be in or just next to some region O fills without being in or just next to all regions O fills. So O can have at once the property of being touched where it fills its region's boundary and the property of not being touched where it fills its region's interior. O need not have parts to have either property. It's not clear, then, why one should go beyond having these properties to having corresponding parts. The move from being touched at the surface of the region to having a surface is no different (or better) than that from O's space's having a left half to O's having one. If O is a true simple, there is no such thing as "that of it which my finger touches." This phrase is tantamount to "the amount of it my finger touches," i.e. how much it touches. How much of the ball my finger touches is either all or some of it. If only some, the ball has parts – those I touch and those I do not. If I touch all of the ball by touching it in one sub-region, the ball as a whole somehow squeezes into that sub-region – the ball is (as Zimmerman has it) "whole in the whole and whole in the parts." Now universals can do this trick. In the area of spacetime an electron inhabits, all of *being negatively charged* is within all of *being negatively charged.* But particulars can't do it. For it requires being "spatially repeated," present as a whole in more than one discrete location in once without time travel, and it is the defining note of a universal to be able to do so. So even answering the question "how much do I touch?" with "all of it" yields an unacceptable result. Like "have you stopped embezzling yet?" this question has no straightforward answer which does not get one in trouble.

This signals that the question has a false presupposition. "Have you stopped embezzling yet?" presupposes that one *has* been embezzling. So answering "yes" or "no" accepts this assumption. This is why the straightforward answers get one in trouble, and honest employees must reject rather than answer this question. "How much do I touch?" presupposes that there *are* amounts or portions of the cueball, and so presupposes that it has parts. So even the straightforward "all of it" treats the ball as parted. This presupposition fails in the case of a simple, be it a Boethian eternal life or a true atom. So the believer in extended simples must reject the "how much" question rather than try to answer it.

This hardly renders belief in extended simples incoherent. For it does not keep one from saying e.g. that the ball itself, not any part of it, is what fills any space the ball fills. Any number of things have no parts and so no "how much." If one asks "how much of redness is in that portrait?," neither "some" nor an "all" whose proper contrast is "some" are proper answers: the question is inappropriate. Redness either just does fill some region of the portrait or just does not. So too, an extended simple either just does fill some region of time or it does not. The question "how much of it is there?," like that about redness in a portrait, just does not arise. This likeness to a universal is not problematic, for it rests on their common trait of not having parts rather than on a trait distinctive of universals.

Tops, bottoms and halves may provide still more challenges. But if we have met these, we may reasonably hope to be able to meet others. Tentatively, then, we can judge an account of region-filling based on overflow as coherent. Now the concepts *region*, *part* and *filling* are mereological and topological, not specifically spatial or temporal. So (1)-(6) apply to both spatial and temporal things. As (1) and (6) mention no parts of A, they let simples overflow regions.

For Boethius, God's life is in each part of time, but does not have a distinct part in any part of time. So I suggest that for Boethius, God's life fills any region of time primarily, overflowing each part of it. If our account of these concepts has been coherent, what Boethius claims seems coherent so far. If things *can* fill their regions this way, then perhaps what Boethius says about an eternal life can be true. I hope to have suggested that the existence of genuine extended atoms of some sort is at least conceptually possible.[4]

NOTES

[1] Boethius, *Philosophiae Consolationis* V, prose 6, in (1926), 400, ll. 9-12, 25-28. My translation.

[2] For support for such a reading, see Brian Leftow (1991, pp. 113-119). For challenges, see e.g. Richard Sorabji (1983, pp. 98-130), and Katherin Rogers (1994, pp. 1-16).

[3] Boethius might also mean that God's life fills regions of non-temporal eternity. But the mereological part of his view is the same either way, so I speak of time to simplify exposition.

[4] My thanks to Trenton Merricks, Philip Quinn, Eleonore Stump and Dean Zimmerman for comments on earlier versions of this material. Thanks too to the Center for Philosophy of Religion at the University of Notre Dame, on whose money and in whose genial company I wrote this paper's first version.

REFERENCES

Boethius (1926), *Philosophiae Consolationis* in H. F. Stewart and E. K. Rand, (eds.), *Boethius: The Theological Tractates*.

Leftow, B. (1991), *Time and Eternity* (Ithaca, NY: Cornell University Press).

Rogers, K. (1994), "Eternity Has No Duration," *Religion Studies* 30: 1-16.

Sorabji, R. (1983), *Time, Creation and the Continuum* (Ithaca, NY: Cornell University Press).

Zimmerman, D. (1996), "Indivisible Parts and Extended Objects," *The Monist* 79: 148-180.

18
TWO PROBLEMS WITH KNOWING THE FUTURE

David Hunt
Whittier College

Knowledge is a good thing. But might it be one of those good things of which one can have too much? Perhaps so, if the knowledge in question is *fore*knowledge. Two concerns over excess foreknowledge have attracted particular attention in the literature. One is that holding a belief about how the future will turn out thereby excludes that portion of the future from being an object of the believer's own agency. The idea is that agency is a matter of *intentional* action, and such action – including activities, like deliberating and deciding, which are oriented toward the formation of intentions – is stultified by an antecedent belief about what one will do. If this is indeed the case, and impotence is an unacceptable price to pay for prescience, it is evidently possible for someone to have more knowledge than is good for him.[1] The other concern regarding foreknowledge is that there are circumstances in which it is incompatible with libertarian freedom. The circumstances in question are those in which the knower is *infallible*. The problem is that infallible knowledge rules out even the *possibility* of someone acting differently than he is known to act, while infallible *fore*knowledge rules out this possibility *in advance*. The latter, however, negates a crucial requirement of the libertarian conception of freedom; so if it is better that the universe contain at least some (libertarian) freedom than that it contain an infallible knower who knows absolutely everything even before it happens, there are conditions under which someone might have more knowledge than it is good for anyone to have.[2]

Let us call the first of these concerns the "Problem of Agency" and the second the "Problem of Freedom." Assuming for the moment that both these concerns are legitimate (an assumption that will be questioned in due course), which concern is more serious? There is no simple answer to this question. The Problem of Freedom is at least potentially the more serious of the two, since an infallible foreknower curtails the (libertarian) freedom of anyone whose actions he foreknows, whereas the only intentionality that can be thwarted by such knowledge is his own (no matter what he foreknows

L. N. Oaklander (ed.), The Importance of Time, 207–223.
 Originally published in *The American Philosophical Quarterly*, 34 (1997), 273–285.

about the actions of others). But in two other respects the Problem of Agency might appear the more formidable. In the first place, a loss of intentional agency would represent an incontestable change in someone's powers, whereas the denial of libertarian freedom might mean no change at all (if, as its critics maintain, such freedom is chimerical in any case). In the second place, attempts to predict the future are far more likely to provoke the Problem of Agency, since its triggering conditions operate at a lower threshold: so long as a potential agent *thinks* that he knows what he will later do, the fact that his belief is fallible, irrational, or even false (any of which would neutralize the Problem of Freedom) is insufficient to mitigate the Problem of Agency. It is, for example, a problem we can imagine arising for human knowers (Oedipus, Macbeth, Billy Pilgrim).[3]

However one assesses the relative seriousness of these two problems, both arise in their strongest form when the foreknower is not a human agent but the theistic God, a being understood to be essentially omniscient, unerring, and sempiternal. If such a God exists and the two problems are genuine, then no one is ever (libertarianly) free, and the most powerful being in the universe is unable to make intentional use of His power. Because the theistic hypothesis makes the central issues so stark, it provides a particularly instructive context in which to examine these two problems, even if one's main interest lies in their implications for human belief and agency. It is not surprising, then, that this context is regularly invoked in discussions not only of the Problem of Freedom (where it is perhaps essential), but of the Problem of Agency as well. The fact that my chief interest in the present paper lies in the logical relationship between these two problems makes it particularly useful to locate them within the theistic environment in which both problems thrive. But this is not an essay in philosophical theology, and references to the Divine Mind will be limited to those which appear to illumine the human situation.[4]

The possibility that there might even *be* an interesting relationship between the Problem of Freedom and the Problem of Agency is one that goes virtually unrecognized in the literature. It is not hard to see why this is so: despite the fact that both problems involve a threat that foreknowledge poses for agency, the brief comparison of these problems on the preceding page is sufficient to indicate the great differences in the range of that threat (the knower v. others), the aspect of agency toward which the threat is directed (intentionality v. metaphysical freedom), and the constituent of foreknowledge that is chiefly responsible for generating the threat (belief v. infallible knowledge). These differences in the logic of the two problems help explain why they are invariably treated as distinct issues. My contention, however, is that these differences mask a more fundamental similarity, and that recognizing this similarity is the first step toward arriving at a proper estimation of these two problems. I begin with the Problem of Agency.

I

Why should one think that foreknowledge engenders a Problem of Agency in the first place? Perhaps it just seems obvious that it does; and admittedly there is a certain intuitive force to the idea that such agential phenomena as intending, deciding, and deliberating are rendered problematic by prior beliefs about what one will do. But it is far from clear just what this problem is, if indeed the intuition that there *is* a problem is even trustworthy. The following argument is supposed to demonstrate that this intuition is in fact sound. It is not the only such argument (two others will be mentioned shortly for sake of contrast); but it is the most important for the purposes of this paper.

There are two fundamental assumptions undergirding this argument. One is that the intentional stance required by genuine agency will be frustrated whenever the agent regards her performing a particular action on a particular occasion as settled or fixed. The second is that a self-reflective agent cannot hold a belief about what she will do at some later time without regarding her performing that action on that occasion as settled or fixed. Given these two assumptions, theistic agency in particular would appear to be self-contradictory. Either God will perform a token of a particular action-type on a particular occasion, or He will not; being omniscient, moreover, He knows which it will be. But then His performing (or not performing) it must be regarded by Him as settled or fixed, leaving Him unable to assume with respect to it the intentional stance of a genuine agent. And the same result will of course follow for any beliefs which a *non*omniscient agent is intemperate enough to form about her own future actions.[5]

Since any argument designed to legitimate the Problem of Agency must show how the gap between antecedent belief and agential impotence is to be spanned, we can distinguish such arguments according to the candidates they put forward for (b) in the series

(a) S believes at t_0 that P: *I will A at* t_1.
(b) ?
(c) S cannot at t_0 engage in acquiring the intention to A at t_1.

Now the essential feature of the argument just adumbrated is the role played by the belief that the future is closed: the agent's belief that she will A at t_1 gives rise to the further belief that the future is closed with respect to her A-ing (or not A-ing) at t_1, and it is this latter belief which makes it impossible for her to engage in decision-making or intention-acquisition with respect to A-ing at t_1. The linking step in this first argument, then, is

(b_1) S believes at t_0 that Q: *my future is now closed with respect to A-ing at* t_1.

Because the erstwhile agent is incapacitated by her own *belief* that the future is closed, let us call this argument the "Doxastic Argument." Before examining the argument more closely, it is worth contrasting its way of connecting (a) and (c) with two others.

The first alternative is to justify the connection by closely identifying the object of belief with the object of intention. Given this conflation of believing and intending, one might then argue that belief pre-empts intention-acquisition (deliberation, choice) because an agent cannot acquire (through the latter) what she already possesses (through the former). Let us call this the "Conflation Argument," and identify its candidate for (b) as

(b_2) S has at t_0 a propositional attitude whose object is P: *I will A at* t_1.

(b_2) is true in virtue of (a); but given the truth of (b_2), it might be thought that a necessary condition for forming the intention to A at t_1 can no longer be satisfied, since the formation of this intention involves acquisition of the very propositional object which (b_2) asserts to be already in S's possession. It is this kind of argument which appears to lie behind, e.g., Carl Ginet's claim that "[t]o believe that someone already knows what his future attempted action will be is to refuse to regard anything he does as having the point necessary to its being his *deciding* that he will attempt the action" (Ginet (1962), p. 52). Here decision-making is ruled out by the antecedent belief alone, regardless of what other beliefs the agent may have – in particular, regardless of whether these include (b_1), the belief that the future is closed.

Unfortunately, the Conflation Argument runs into trouble with its main assumption that the objects of intention may be assimilated to the objects of belief. Intending and believing must be distinguished in some way, and it is hard to think how else this can be done except in terms of their objects or the attitudes directed toward those objects. If the difference lies in the propositional objects, then the assumption is simply false: the object of belief is the declarative proposition *I will do such-and-such*, while the object of intention is the optative proposition *would that I might do such-and-such*. The difference in mood undermines the claim that the objects are the same in such a way that someone who possesses the one cannot go on to acquire the other. But if the difference lies in the propositional attitudes, the fact (if it be a fact) that these attitudes are directed toward identical objects becomes irrelevant to the Conflation Argument's success; for having an attitude of belief toward the proposition *I will A at t_2* does not by itself settle the question whether one has or will come to have the quite different attitude of intention toward the same proposition. There is therefore no reason to think (c) true on the basis of (b_2). Human agents who anticipate their own actions might mistakenly conflate believing with intending and find their agency hamstrung as a result – but they might also recognize and avoid the mistake, and an infallible deity would surely do so.[6]

The second way of justifying the connection between antecedent belief and agential impotence without bringing in an explicit belief that the future is closed is to point to the conditions under which the agent might come to have a belief about his own future actions. Richard Taylor, for example, identifies these conditions as follows: "There seem, in fact, to be only these two ways in which one could know what he is going to do; namely, by *inferring* what he is going to do, or by *deciding* what he is going to do. In neither case can one deliberate about what he is going to do" (Taylor (1964), p. 75). In other words, (a) will be true only if

(b_3) It is the case at t_0 either that S has already acquired the intention to A at t_1 *or* that S has inferred her A-ing at t_1 from other things she believes at t_0.

But (b_3) appears to entail (c): for *inference* rests on propositions whose truth is taken to be independent of what remains open for deliberation, while *intention* makes further deliberation impossible – at least until such time as the intention is withdrawn, at which point the foreknowledge based on that intention must also go.

This argument – call it the "Conditions Argument" – also founders on its main assumption. In the first place, there is no reason to think that *God* must acquire His knowledge of the future in either of these two ways; indeed, God is commonly thought of as knowing the future in the way that we know the present: not by deciding how to manipulate it, or by inference from other facts, but through a direct apprehension. In the second place, it is far from clear that *human* foreknowledge is confined to these two methods (assuming that time machines, tachyonic radios, crystal balls, or the like are conceptually possible); and even if it were so confined, fore*belief* is certainly not, and it is belief (rather than the other constituents of knowledge) which is relevant to the Problem of Agency. It is, for example, an empirical fact that people have acquired beliefs about their own future actions by reading tea leaves, frequenting fortunetellers, and having "precognitive" experiences; nor is there reason to doubt that such beliefs could be acquired as the result of hypnotism, a blow on the head, or the ministrations of a demented neurophysiologist. Because none of these beliefs is acquired under either of the conditions mentioned by Taylor, there is no reason (so far as the Conditions Argument goes) to think any of them acquired under conditions which nullify subsequent deliberation.

II

So much for the main alternatives to the Doxastic Argument. Since both suffer crippling defects, the Problem of Agency appears to depend for its cogency on the Doxastic Argument. Thus motivated, let us return to that argument, considering first the move from (b_1) to (c) and its assumption that a potential agent is prevented from embarking on the path to intentional agency by the belief that her options are closed. The formation and maintenance of intentions may, of course, be thwarted in many ways: through death, ingestion of a desire-altering drug, and the like. But in a normally functioning rational agent it may also be thwarted by its lack of "fit" with other elements in the agent's psychological economy. In particular, there is a practical inconsistency – an egregious lack of "fit" – between one's acquiring the intention to perform a token of a particular action-type and the prior belief that one's performing or not performing a token of that type is no longer an open issue. For if the action is believed to be already necessary, the intention to perform it must be judged superfluous; if the action is believed to be already impossible, the intention to perform it must be judged ineffective. In either case it is futile to engage in activities, such as deliberation and decision-making, whose whole point is the acquisition of intentions. Any rational agent who believed her options to be closed would therefore refrain from such activities (a failure to do so perhaps even being regarded as a mark of cognitive dysfunction).[7] This claim may in fact admit of unusual exceptions; but as the exceptional cases yield unimpressive examples of agency, the argument is not best addressed by exploiting these exceptions, and I propose that the claim be accepted.

That leaves the move from (a) to (b_1). Why exactly must the belief that one's future is closed with respect to A-ing follow from the belief that one will A? One explanation might advert to the brute fact that believers are "wired" in such a way that the one belief cannot be held without the other. But this account is of dubious value on two grounds. In the first place, it is doubtful that anyone putting forward this claim is really in possession of the relevant information about how believers are "wired." And in the second place, even if such information were available, so long as this stricture on belief is *merely* a brute fact, unsupported by any rationale for the stricture, it could have no bearing on agents with different "wiring." This includes, most notably, God; but also human agents in some futuristic scenario who have been "rewired" by a neural technician (an operation which might appeal, e.g., to psychics who find their attempts at intentional agency hamstrung by precognitions of their own actions).

A better explanation is that this move, like the one from (b_1) to (c), is required by considerations of rational mental functioning. This is the approach taken by Tomis Kapitan, the principal proponent of the Doxastic Argument. His position turns on the claim that an agent's believing an action to be an open alternative for her amounts to her believing it to be *contingent* in an appropriate sense, where the appropriate sense is contingency relative to the agent's own (accessed) beliefs. If this claim is accepted, the Doxastic Argument goes through. For suppose that (a) is true, that is, that S believes at t_0 that P: *I will A at t_1*. Now clearly P is not contingent relative to itself; moreover, we can assume that a rational agent would not believe P to be contingent relative to itself (if necessary, eschewing such a belief can be made a requirement for rational agency). But then if S believes that P, it follows (if she is self-reflective as well as rational) that she will not regard P as contingent relative to her own beliefs. According to Kapitan, she will therefore regard her future with respect to A-ing at t_1 as closed, just as (b_1) asserts.[8]

Is this account of agential openness in fact acceptable? Certainly much about it has an air of plausibility. It is plausible, for example, to construe openness as a kind of contingency; and given that the openness of an action-type for an agent can vary with circumstances (changes in the agent's abilities, in external opportunities, and so on), it is plausible to construe such openness as a relative modality, namely, contingency relative to some set of propositions $\{\Pi\}$. Furthermore, given that we are concerned here with a doxastic stricture on an agent's intentional capacities, it is surely correct to construe $\{\Pi\}$ as consisting wholly of propositions believed by the agent. But *which* propositions believed by the agent? More specifically, why suppose that the propositions in question will include *P* (rather than limiting the relevant propositions to those, if any, which S believes to set forth the *deterministic causes* of her action)? The answer to this question is far from obvious.

If the answer is to be found in the demands of rationality, we should be able to identify some benefit accruing to the agent from the belief that her own foreknown actions are closed, so that even God and psychic self-predictors would be worse off without this belief. There are two sorts of contexts in which this might be the case. In the first place, planning for the future would be impossibly complex if in the course of practical reasoning we had to regard as open all those aspects of the future which (so far as we know) actually are open. Thus it is essential that we make certain simplifying assumptions about the future, which we can then draw upon in attempting to reach decisions about other aspects of the future. (Perhaps God too would find such assumptions useful.) These include not only assumptions about the conditions in which we will be acting ("Assuming that it rains tomorrow..."), but also assumptions about our own actions ("Given that I accept the offer..."). Such assumptions may range from the purely hypothetical ("Were I to accept the offer...") to the quite definite ("Since I am going to accept the offer..."); in either case the exigencies of planning may lead one to regard the issue (at least provisionally) as *closed* for purposes of further practical reasoning.[9]

But clearly this sense in which the future might be regarded as closed does nothing to buttress the Doxastic Argument. Assumptions on the hypothetical end of the spectrum do not support a *belief* about the future, and the sense of closure they give rise to is purely heuristic. Assumptions on the other end of the spectrum, which reflect definite decisions, are also useless to the Problem of Agency; for while they may rise to the level of belief ("I will accept the offer"), such beliefs about the agent's future come with (and depend upon) the agent's decision about her future, rather than pre-empting it; and while the sense of closure accompanying this belief is perhaps more than purely heuristic, it is a product of the agent's successful exercise of practical reasoning and is thus supportive of agency rather than subversive of it.

The second context in which the agent would benefit from the belief that the future is closed is where the future really is closed, independent of the agent's own decisions or heuristic assumptions; for if there *are* no alternatives to S's A-ing at t_1, the *belief* that there are no alternatives to her A-ing at t_1 will have the beneficial effect of diverting S away from unpromising lines of practical reasoning. The chances that such a belief will be held in a context in which it is true are of course increased if the agent has good *reasons* to hold the belief, and this suggests that the Doxastic Argument's assertion of a necessary connection between the belief that one will A at t_1 and the belief that one's future is closed with respect to A-ing at t_1 is simply the claim that the former belief constitutes a rationally compelling reason for the latter. This seems to be just the sort of connection required by the Doxastic Argument; for if S's belief that she will A at t_1 does *not* make it rational for her to believe that her future is closed with respect to A-ing at t_1, it is hard to see in what way these beliefs *could* be connected so as to be applicable to God, or even support some interesting generalization about human agency. This will of course restrict any generalization to *rational* agency; but this is where real interest in the Problem of Agency lies in any case.

For the Doxastic Argument to succeed, then, it must be the case that any (rational and self-reflective) agent who believes at t_0 that

(P) I will A at t_1

will also believe at t_0 that no alternatives to P are then available to her. She will believe, that is, that

(Q) $\Box_0 P$,

where, "$\Box_0$" symbolizes that P is true in all worlds which the agent has the power to bring about at t_0. So understood, the central question regarding the Doxastic Argument appears to be whether $P \supset Q$ – whether, that is $P \supset \Box_0 P$. For if it is true that $P \supset \Box_0 P$ – and true in so straightforward a manner that only an imbecile would fail to see it – then any rational agent who believes P will also believe $\Box_0 P$; but if it is false that $P \supset \Box_0 P$, it is a total mystery why every rational agent would nevertheless believe $\Box_0 P$ upon believing P, and do so as a requirement of being rational.

III

In considering whether it is indeed straightforwardly true that $P \supset \Box_0 P$, it will be helpful to look at some of the ways in which one might try to move from the proposition that one will do such-and-such to the proposition that the future is closed with respect to such-and-suching. Most are obvious nonstarters for the purpose at hand. For example, the conjunction of P with

(R) My future is now closed with respect to my action(s) at t_1 (whatever they might be)

implies $\Box_0 P$; but R is simply trumped up to achieve this result, and there is absolutely no reason to think that every rational agent would accept it (though it is easy to think of situations in which some might). Again, the conjunction of P with

(S) My future is causally determined by my past

implies $\Box_0 P$; but the considerations underlying S are so far from straightforward that acceptance of S can hardly be made a requirement of rationality, even if S happens to be true. In contrast to these approaches, the idea that $\Box_0 P$ follows straightforwardly from P (in such a way that the Problem of Agency is unavoidable for rational agents) is characteristic of arguments for *fatalism*. Let's look briefly at three such arguments, beginning with the simplest (and thus the one most likely to have a universal appeal).

The first "argument" (it's hard to dignify it this way) justifies the inference from P to $\Box_0 P$ on the grounds that, "if I am going to A at t_1, then I am going to A at t_1." How exactly this is supposed to support the inference is far from clear, but it may have something to do with the fact that the conditional offered as justification is a tautology and thus itself necessary. Suppose the necessity could be parsed this way: "If I am going to A at t_1, then, necessarily, I am going to A at t_1." Then the argument would go like this: P, P $\supset$ necessarily P, $\therefore$ necessarily P. This argument has the virtue of being valid, but the vice of misrepresenting the tautology, which must instead be rendered, "Necessarily, if I am going to A at t_1, then I am going to A at t_1." The argument then comes to this: P, necessarily (P $\supset$ P), $\therefore$ necessarily P. But this involves an egregious modal fallacy ("Sleigh's fallacy," as it has been called). Because it is so egregious, let us call the position based on this fallacy "Naive Fatalism." Insofar as the Doxastic Argument presupposes that we all reason like Naive Fatalists, it errs on a matter of fact (since we do not all reason this way) and disqualifies itself as a stricture on *rational* agency (since it errs in ways that rational agents can, and should, avoid).[12]

The next approach regards the intuitions underlying the preceding argument as basically sound, but in need of a more philosophically sophisticated expression. Let us therefore call this position "Sophisticated Fatalism." The idea is that what makes it unavoidable that I am going to A at t_1 is that it is already true that I am going to A at t_1, and what is already true is necessary in the sense that no one (not me, not even God) can do anything about it now. This allows the argument of the Naive Fatalist to be restated so that it appears to avoid the fallacy. The new argument goes like this. If P, then

(T) It was true before I was born that P.

And T of course entails P. But if T, then $\Box_0 T$ (since I can't do anything now about what was the case before I was even born). So we get the valid argument : $\Box_0 T$, $\Box$ (T $\supset$ P), $\therefore$ $\Box_0 P$. Unfortunately for Sophisticated Fatalism, escape from its Naive cousin's invalidity comes at the price of unsoundness. T is a classic example of what has come to be called a "soft fact" about the past, and only "hard facts" are unavoidable constituents of the past. According to William Ockham, who first deployed the distinction to argue against fatalism, soft facts "are about the present or about the past as regards their wording only and are equivalently about the future" (Ockham (1969), pp. 46-47), while hard facts are really (and not just verbally) about the past or present. "It was true before I was born that I will A at t_1" is about as "soft" as any fact about the past can be. Its truth depends entirely on what happens at t_1; it does not "harden" as a fact until t_1 arrives. "$\Box_0 T$" is therefore unacceptable as a premise: it doesn't follow from P unless there is some independent reason to think that $\Box_0 P$, in which case it would be pointless to employ it as a reason for believing that $\Box_0 P$. So Sophisticated Fatalism won't do as a basis for the Problem of Agency.[11]

The problem with T can be remedied by finding some other proposition which entails P and is also unavoidably necessary at t_0. Classical theism provides such a proposition:

(U) God believed before I was born that P.

The argument for "Theological Fatalism," as we might dub it, goes like this: $\Box_0 U$, $\Box (U \supset P)$, $\therefore \Box_0 P$. This argument, like the one for Sophisticated Fatalism, is valid; unlike the argument for Sophisticated Fatalism, however, it may well be sound. Because the beliefs of the theistic God are infallible, "$\Box (U \supset P)$" must be premised as true; and because the occurrence of a belief at a time, unlike the truth of a proposition at a time, is surely a "hard" (rather than a "soft") fact about that time, it would appear that "$\Box_0 U$" is true as well (assuming that the theistic God exists). Human beliefs, after all, are no less "hard" than other historical events. While such beliefs may certainly figure in soft facts about the past – e.g., "In 1988, three years before it actually happened, Higgenbotham believed that the USSR would break up," which is soft relative to 1989 – the simple assertion that a belief occurred at a time is always a hard fact about the past – e.g., "In 1988 Higgenbotham believed that the USSR would break up," which is hard relative to 1989. There is no apparent reason why divine beliefs (assuming that they occur in time) should be different from human beliefs in this respect, however different they may be in other respects.[12]

Of these three fatalistic arguments, only the one for Theological Fatalism avoids obvious error. This is not to say that it is in fact successful: the argument does assume, after all, that the theistic God (or some other infallible foreknower) exists; furthermore, there is continuing controversy over whether (U) is indeed a hard fact prior to t_1 (in contrast with (T), which is certainly *not* a hard fact). But at least Theological Fatalism, unlike its Naive and Sophisticated cousins, is not a nonstarter.[13] It is therefore the only one of the three that might be rational for someone to accept. Of course, any of the fatalistic arguments might be accepted by an *ir*rational agent, or by a rational agent suffering a momentary and regrettable loss of reason. But these are the uninteresting cases of the Problem of Agency. It is the Theological Fatalist alone who faces the Problem of Agency in its only interesting form.[14]

Perhaps there are other arguments that do a comparable or superior job of explaining why $P \supset Q$ – I certainly haven't canvassed all the possible arguments here. But the fact of the matter is that the friends of the Problem of Agency have failed to identify these alternative arguments. In the face of this failure, I make so bold as to propose that there is no (interesting) Problem of Agency unless Theological Fatalism is true. The onus is on the person denying this claim to come forward with the alternative argument by which the Problem of Agency might be generated for a rational agent who holds a belief about her own future actions.

IV

We began with two problems supposedly generated by knowledge of the future. Our quest to understand the Problem of Agency has led us to the Doxastic Argument and the grounds for supposing that the belief that I will A at t_1 makes it rationally compelling for me to believe as well that my A-ing at t_1 is now unavoidably necessary. It is hard to imagine what these grounds might be apart from something like Theological Fatalism. If this assessment is correct, consideration of the Problem of Agency has brought us face-to-face with the second of our two foreknowledge problems; for the Problem of Freedom is nothing other than the predicament posed by Theological Fatalism, with the addition of a premise connecting libertarian freedom with the ability to do otherwise (holding the actual past fixed).

There is a close parallel, then, between the two foreknowledge problems: while the Problem of Agency begins with the *belief* that one will do such-and-such, moves from there to the *belief* that the future is closed with respect to such-and-suching, and concludes from this belief that *intentional agency* is impossible, the Problem of Freedom begins with the *fact* that one will do such-and-such, moves from there to the *fact* that the future is closed with respect to such-and-suching, and concludes from this fact that *libertarian freedom* is impossible. Moreover, because the beliefs that figure in the Problem of Agency are connected only by the ties of rationality, they presuppose the connections between facts that figure in the Problem of Freedom. What must be *true* for the latter to be genuine, must be *believed* true for the former to arise, and must be *both* believed *and* true for the former to be completely rational. In short, the first problem may be regarded as a doxastic version of the second problem.

This means that a number of assumptions that are well-nigh universal in the contemporary discussion of these two problems are instead mistaken. These assumptions are: (1) that the Problem of Agency is logically independent of the Problem of Freedom; (2) that the Problem of Agency imposes a general stricture on (rational) agency; (3) that it is only the libertarian conception of (free) agency that is threatened by the Problem of Freedom; and (4) that the Problem of Agency is most serious for an omniscient foreknower like the theistic God. I will now summarize what is mistaken in each of these assumptions.

(1) The Problem of Agency is parasitic on the Problem of Freedom. This point is already implicit in the foregoing analysis, but is worth making explicit. The Problem of Agency presupposes the presumptive agent's acceptance of a fatalistic argument; but only the argument for Theological Fatalism stands any chance of being sound; therefore the Problem of Agency is ineliminable for a rational agent only if the argument for Theological Fatalism is sound. Since the latter argument sets forth the Problem of Freedom, the Problem of Agency (in its only interesting form) is dependent on the Problem of Freedom. This means that the independent treatment of these problems in the contemporary debate is misguided. The Problem of Agency does not exist in splendid isolation from the Problem of Freedom, but depends for its cogency on the Problem of Freedom. The two problems are really one, and the resolution of this joint problem depends on current investigations of the Problem of Freedom rather than the Problem of Agency.

(2) The Problem of Agency cannot be regarded as a stricture on rational agency *per se*. Whether it makes agency problematic in any given case depends on whether the erstwhile agent infers Q from P; and whether it makes *rational* agency problematic depends on whether it is *rational* for the agent to infer Q from P. At worst, it might always be fallacious to infer Q from P, in which case the Problem of Agency would arise only in certain cases of *ir*rational agency. At best, there might be a fatalistic argument acceptance of which would not constitute an obvious blunder, in which case rational agency would be jeopardized only for aspiring agents who accept the argument's premises. Despite their intuitive appeal, such expressions of the Problem

of Agency as the maxim, "One can't deliberate over what one already knows is going to happen," do not impose a general stricture on rational agency (though they could well constrain the agency of theists who are also Theological Fatalists).

(3) The Problem of Freedom does more than annul the libertarian freedom of actions that are infallibly foreknown; it raises the Problem of Agency as well. If the argument for Theological Fatalism is sound, it may be rational to accept it; and if it is rational to accept it, then anyone who does accept it and holds a belief about her own future actions has a rational basis for reaching a fatalistic conclusion about those actions, and for abandoning agential initiatives directed toward those actions. Thus it is not just a particular and controversial conception of *free* agency that is at stake in the Problem of Freedom, but the agency *simpliciter* (libertarian or otherwise) of forebelievers unfortunate enough to realize that the argument for Theological Fatalism works. Given the assumptions underlying that argument, this will be a particular occupational hazard of rational and self-reflective theists – among whom must be numbered God, who is a rational and self-reflective theist if anyone is. Among the actions God foreknows are His own future actions; so these will be just as fated as human actions, if the argument for Theological Fatalism is sound. But if the future doings of an omniscient being are unavoidably necessary, He must know that they are unavoidably necessary (otherwise He wouldn't be omniscient). So His actions *in toto* are not only infected by the Problem of Freedom, but by the Problem of Agency as well.

(4) Despite what has just been said, the Problem of Agency is not necessarily more serious for the theistic God than it is for human beings. It is true that God, in virtue of His omniscience, will know in advance all His own future doings, and so will have *more* beliefs about what He will do than will a limited human knower about what she will do; but this will translate into a greater Problem of Agency only if there is a sound argument for fatalism. If there is, then an omniscient being will of course be aware of that argument and grasp its fatalistic implications, with agential sclerosis the necessary consequence – divine omniscience, as Tomis Kapitan is fond of claiming, will entail divine impotence. But if there is no sound argument for fatalism,

> it won't matter that God knows in advance everything He is going to do, since He will resist inferring unavoidability from the mere fact that an action is going to take place. A nonomniscient being might endorse a mistaken inference, thereby jeopardizing her agency in the (presumably small) number of cases where she holds a prior belief about what she will do; but an omniscient being would never endorse a mistake.

If I am right, the real action is to be found in the debates surrounding the Problem of Freedom, with the Problem of Agency (now in a stripped-down form) crucially dependent on the outcome of those debates. This does, however, leave an important question unresolved. Why is the Problem of Agency so viscerally compelling, apart from the considerations brought out in the Problem of Freedom? Isn't the immediate claim the Problem of Agency makes on our intuitions incompatible with its really resting on an *argument* connecting P with Q? This is perhaps a psychological rather than a philosophical question, though it is worth noting that Naive Fatalism is also immediate and visceral in its appeal. I suspect that the Problem of Agency, as it actually arises for most of us, rests less on Theological Fatalism or even Sophisticated Fatalism, and more on the seductive draw of Naive Fatalism. Consider a soldier in a foxhole who reasons as follows: either there is a bullet with my name on it, or there isn't; if there is, it is useless to keep my head down; if there isn't, it is unnecessary to keep my head down; therefore, it is pointless to keep my head down. It is quite possible to identify the fallacy in this sort of argument only to find oneself falling back under its spell five minutes later – indeed, I suspect that this is a rather common experience (at least among people capable of picking out the fallacy in the first place). This being the case, if the Problem of Agency is simply a doxastic form of fatalism, as I have argued, its intuitive appeal is just what one would expect, given the larger phenomenon of fatalism's deceptive hold over our judgment.

NOTES

[1] Some of the sources in which this problem receives sympathetic treatment are: Ginet (1962); Taylor (1964); La Croix (1976); Denyer (1981); Kapitan (1984); Dennett (1984); Kapitan (1986); and Strawson (1986). No one claims that agency is threatened by the subject's *nonoccurrent* or *unaccessed* beliefs; consequently I assume in what follows that it is only *occurrent* beliefs that are in question.

[2] The seminal source for this problem in the recent discussion is Nelson Pike (1965). The books and articles generated by Pike's essay are too numerous even to begin to list.

[3] While the Problem of Freedom in the narrow sense cannot be generated by human foreknowledge, there may well be problems in the neighborhood of this one that are so

generated. If I arrive from the future in a time machine and know what you are going to do because I just saw you do it, are there adverse consequences for your putative libertarian freedom? Whatever the answer to this question, it is certainly odd that it should depend on the fact that my knowledge fails to be absolutely *infallible*. If this intuition is sound, it may be that the Problem of Freedom (or at least a very close cousin) can be found in cases of human foreknowledge (as well as divine).

[4] I have addressed the theological implications of the Problem of Agency in (1992), and (1993), while in (1995a), I defend a novel conception of divine knowledge which permits solutions to both the Problem of Agency and the Problem of Freedom.

[5] This summary statement of the argument is closest to the formulation given it by Tomis Kapitan in two recent articles: (1990) and (1991). But the basic idea may also be found in Denyer (1982, p. 50); and David Basinger (1986, p. 171).

[6] This response to the Conflation Argument is much influenced by Alvin Goldman's treatment of the issue in (1970). He introduces the distinction between optative and declarative propositions on p. 102, and applies the distinction to the Problem of Agency on pp. 176-194.

[7] See Kapitan's discussion of the "principle of least effort" in his (1991, pp. 109-110).

[8] Openness as contingency relative to the agent's occurrent beliefs becomes an explicit and established part of Kapitan's analysis of the Problem of Agency with his (1986).

[9] For further discussion of this practical sense of closure, see Bratman, (1987). Bratman argues that intending to A does not necessarily imply that one believes that one will A (though it does imply *not* believing that one will *not* A); it does, however, involve a *commitment* to A-ing, and this in turn plays a role in one's further plans, forming a background against which further intendings are acquired. Having intended to A, thus being committed to A-ing, makes it inconsistent from an *internal* perspective for one not to A, though it is not inconsistent from an *external* perspective (since one presumably has the *power* not to A). Though reconsideration is always possible, there is a certain presumption against reconsideration and in favor of stability – otherwise provisional intentions could not be used as standards for the relevance and admissibility of further options, as his "planning" theory requires.

[10] Perhaps the best-known defense of Naive Fatalism by a contemporary philosopher is Taylor's in the chapter on "Fate" (1963).

[11] Recent versions of Sophisticated Fatalism may be found in Ryle, (1954), and Cahn, (1967).

[12] Some contributors to the modern debate, following Ockham's own lead, have argued that U as well as T is a soft fact about the past. The seminal articles for this "Ockhamist" critique of Theological Fatalism are Saunders (1966), and McCord Adams (1967). In (1995b), I argue that the Ockhamist position might be plausible if divine beliefs are understood nonoccurrently.

[13] For the superiority of theological over logical fatalism, see Widerker (1989). A good discussion may also be found in the first part of Pike's "Theological Fatalism and Prior Truth" (unpublished manuscript).

[14] The arguments originally offered by Tomis Kapitan appear to presuppose (or at least flirt with) Naive Fatalism. More recently, however, he has suggested that the Doxastic Argument could rest equally well on Theological Fatalism. For a discussion of this issue, see Kapitan (1994), and my (1996).

REFERENCES

Basinger, D. (1986), "Omniscience and Deliberation: A Response to Reichenbach," *International Journal for Philosophy of Religion* 20: 169-72.

Bratman, M. E. (1987), *Intentions, Plans, and Practical Reason* (Cambridge, MA: Harvard University Press).
Burton, R. G. (1982), "Choice," *Philosophy and Phenomenological Research* 42: 581-586.
Cahn, S. (1967), *Fate, Logic, and Time* (New Haven: Yale University Press).
Dennett, D. C. (1984), *Elbow Room: The Varieties of Free Will Worth Wanting* (Cambridge: MIT Press).
Denyer, N. (1981), *Time, Action and Necessity* (London: Duckworth).
Ginet, C. (1962), "Can and Will Be Caused?" *Philosophical Review* 71: 49-55.
Goldman, A. (1970), *A Theory of Human Action* (Princeton: Princeton University Press).
Hunt, D. (1992), "Omniprescient Agency," *Religious Studies* 28: 351-369.
Hunt, D. (1993), "Divine Providence and Simple Foreknowledge," *Faith and Philosophy* 10: 396-416.
Hunt, D. (1995a), "Dispositional Omniscience," *Philosophical Studies* 80: 243-278.
Hunt, D. (1995b), "Does Theological Fatalism Rest on an Equivocation?" *American Philosophical Quarterly* 32: 153-165.
Hunt, D. (1996), "The Compatibility of Omniscience and International Action: A Reply to Tomis Kapitan," *Religious Studies* 32: 49-60.
Kapitan, T. (1984), "Can God Make Up His Mind?" *International Journal for Philosophy of Religion* 15: 37-47.
Kapitan, T. (1986), "Deliberation and the Presumption of Open Alternatives," *Philosophical Quarterly* 36: 230-251.
Kapitan, T. (1990), "Action, Uncertainty, and Divine Impotence," *Analysis* 50: 127-133.
Kapitan, T. (1991), "Agency and Omniscience," *Religious Studies* 27: 105-121.
Kapitan, T. (1994), he Incompatibility of Omniscience and International Action: A Reply to David P. Hunt," *Religious Studies* 30: 55-66.
La Croix, R. (1976), Ómnipresciencе and Divine Determinism," *Religious Studies* 12: 365-381.
McCord, M. (1967), "Is the Existence of God a 'Hard Fact'?" *Philosophical Review* 76: 492-503.
Ockham (1969), in Marilyn McCord Adams and Norman Kretzman (trans. and intro.), *Predestination, God's Foreknowledge, and Future Contingents,* (New York: Century Philosophy Sourcebooks, Appleton-Century-Crofts).
Pike, N. (1965), "Divine Foreknowledge and Voluntary Action," *Philosophical Review* 74: 27-46.
Pike, N. (unpublished), "Theological Fatalism and Prior Truth."
Ryle, G. (1954), "Ít Was To Be," in *Dilemmas* (Cambridge: Cambridge University Press).
Saunders, J. T. (1966), "Of God and Freedom," *Philosophical Review* 75: 219-225.
Strawson, G. (1986), *Freedom and Belief* (Oxford: Clarendon Press).
Taylor, R. (1963), "Fate," in *Metaphysics* (Englewood Cliffs, NJ: Princeton-Hall).
Taylor, R. (1964), "Deliberation and Foreknowledge," *American Philosophical Quarterly* 1: 73-80.
Widerker, D. (1989), "Two Forms of Fatalism," in J. M. Fischer (ed.), *God, Foreknowledge and Freedom* (Stanford: Stanford University Press): 97-110.

Part Five

TIME AND PHYSICAL OBJECTS

19

TIME, SPACE, AND THE NATURE OF PHYSICAL OBJECTS

Ned Markosian
Western Washington University

1. INTRODUCTION

The topic of this paper is a concept that has figured prominently in the history of philosophy, and is probably more important now than it has ever been before. It plays a crucial role in contemporary discussions of the nature of people, the nature of mental phenomena, and ontological theses like materialism (the thesis that only physical objects exist) and physicalism (the thesis that everything that exists is just what a true and complete physics would say exists). In addition, the vast majority of living philosophers begin their ontologies with a category for objects that fall under this concept. I refer, of course, to the concept of a physical object.

Despite its obvious importance in philosophical discussions, the concept of a physical object has received surprisingly little attention. In particular, the question "What are physical objects?" i.e., "What is the correct analysis of the concept of a physical object?" has been all but ignored by most philosophers. The purpose of this paper is to address this question. In what follows, I will consider several attempts at answering the question, and give my reasons for preferring one of them over its rivals. The account of physical objects that I will recommend is one that defines physical objects as objects with spatial locations.

2. THE SPATIAL LOCATION ACCOUNT OF PHYSICAL OBJECTS

Hobbes said that "a body is that, which having no dependence on our thought, is coincident or coextended with some part of space" (*De Corpore*, 2.8.1.). If I understand Hobbes correctly then the analysis of the concept of a physical object that I want to recommend is his. For on my view, the best

L. N. Oaklander (ed.), The Importance of Time, 227–241.

Originally published as "What Are Physical Objects" in *Philosophy & Phenomenological Research*, 61 (2000), 375–395.

account of physical objects is this: to say that something is a physical object is to say that it is an object with a spatial location. I'll call this view the Spatial Location Account of Physical Objects.

The intuitive idea behind the Spatial Location Account is this. Objects from all of the different ontological categories – physical objects; non-physical objects like souls, if there are any; propositions; universals; etc. have this much in common: they all exist *in time*. But not all of them exist *in space*. The ones that exist in time and space, i.e., the ones that have spatial locations, are the ones that count as physical objects. Thus souls, if there are any, are objects with temporal locations but without spatial locations. They might interact with objects in space, like our bodies, but even when they do so, they are not themselves located in space.

It should be obvious that the Spatial Location Account requires a sharp distinction between time and the dimensions of space. For on the Spatial Location Account, the defining characteristic of physical objects is spatial location, not temporal location. And if there is a sharp distinction between spatial location and temporal location, as there must be in order for the former to be the defining characteristic of physical objects while the latter is not, then there must also be a sharp distinction between time and the dimensions of space.

Many philosophers believe ours to be a world of four basically similar dimensions, but I don't. On my view, there is a sharp distinction between time and the dimensions of space. I "take tense seriously," as they say. I believe the "3D view" of persisting objects, according to which objects that persist through time do so in virtue of being wholly present at each moment at which they are present, rather than in virtue of having different temporal parts at the different moments. And I believe that there is an important sense in which it is true to say that time passes although no dimension of space does so.[2] So although it may seem to some that requiring a sharp distinction between time and the dimensions of space is a major disadvantage of the Spatial Location Account of Physical Objects, this same feature of the account seems to me like a virtue.

The Spatial Location Account of Physical Objects should not be confused with another view about phsyical objects that also makes reference to space. This is the view that physical objects are objects with spatial *extensions*. The Spatial Extension Account of Physical Objects rules out the possibility of "point-sized" physical objects, such as quarks, but the Spatial Location Account does not. Some people, who don't want to count quarks as physical objects, might see this as a disadvantage of the Spatial Location Account, but I see it as an advantage. I will say a little bit more about quarks, and why I think they should count as physical objects, below. But first, let's consider some alternative accounts of physical objects.

3. SOME IMPLAUSIBLE ACCOUNTS OF PHYSICAL OBJECTS

One natural response to the question "What are physical objects?" is simply that physical objects are the objects studied by physics.[3] I will call this the Physical Theory Account of Physical Objects. A major problem with the Physical Theory Account is that it seems likely to lead to circularity, since it's natural to want to define physics as the study of physical objects and the laws of nature governing them.[4] In fact, I am inclined to say that the expressions "physical object" and "physics" are best understood as being interdefinable. If we knew what either one meant, then we would know what the other means too; but I don't think we can get a really informative definition of either one in terms of the other.

Another natural way of answering the question "What are physical objects?" is to say that physical objects are the objects that can be sensed.[5] But on the face of it there is a grave difficulty with this view: what can be sensed varies from subject to subject. Thus it appears that, on the Sensational Account, the concept of a physical object will turn out to be a relativistic one; a given object might be a physical object relative to a certain dog, say, but a non-physical object relative to a particular human. This kind of result seems undesirable, especially considering the crucial role that the concept of a physical object plays in philosophical controversies.

The Sensational Account can perhaps get around this objection if we understand it to say that physical objects are objects that can be sensed by some sentient being or other. But the view will still face grave difficulties. It's not at all clear which objects can be sensed and which cannot. Many dualists believe that minds, for example, can be sensed by the faculty of "inner sense" – and yet we do not want minds to count as physical. Moreover, it's not clear that any object is in principle incapable of being sensed. It seems conceivable that even the proposition that 2 + 2 = 4 could be sensed by a sufficiently sensitive being.

Considering these difficulties with the Physical Theory Account and the Sensational Account, it might seem that the wisest course would be to take the concept of a physical object as a simple, unanalyzable concept. But I think it would be overly hasty, at this point, to do so. I think that when dealing with such an important concept, we should prefer a more or less plausible analysis that seems to give fairly intuitive results to the position that there is no interesting analysis of the relevant concept. And I think that there are indeed available several more or less plausible analyses of the concept of a physical object, including the Spatial Location Account. Thus I think it would be best to adopt one of those analyses. Let's turn now to some of the rivals of the Spatial Location Account that, like it, seem preferable to the Simple Account of Physical Objects.

4. QUINE'S ACCOUNT OF PHYSICAL OBJECTS

W. V. Quine is one of the few philosophers I know of who explicitly addresses the question "What are physical objects?". In his paper, "Whither Physical Objects?," he says the following:

> ... let us understand a physical object, for a while, simply as the aggregate material content of any portion of space-time, however ragged and discontinuous. (Quine (1976), p. 497)

The reason Quine says "for a while" is that he goes on to give an account of what can constitute the "material content" of any region of space-time, and, on his account, talk about material contents is to be replaced by talk about distributions of states over space-time (1976, p. 499). Thus, on Quine's view, physical objects actually *are* distributions of states over space-time. But there is more. He goes on to analyze both states and space-time regions as constructions out of pure set theory (1976, pp. 499-502). Thus, according to Quine, physical objects are actually distributions of purely set-theoretic constructions over other purely set-theoretic constructions.

I don't know exactly what "distributions" are, according to Quine, but apparently they end up being constructions out of pure set theory as well, for Quine later says, by way of summarizing his account of physical objects,

> Physical objects... evaporated into space-time regions... Finally the regions went over into pure sets... (1976, pp. 502-503)

Thus, according to Quine, physical objects ultimately *are* constructions out of pure set theory.

Now, with all due respect, this just seems wrong. It can't be that physical objects are somehow constructions out of pure set theory, for physical objects are contingent, while set-theoretic constructions (from pure set theory, at least) exist necessarily. Therefore, no physical object is identical to any construction out of pure set theory.

I can think of two main ways Quine might respond to this modal objection. First of all, being no friend of modality, he might simply reject as meaningless such modal talk as the claim that physical objects are contingent while constructions out of pure set theory exist necessarily. But I am a friend of modality, and so this reponse seems untenable to me. certain purely set-theoretic constructions, while existing necessarily,

An alternative response to the objection that Quine might make is that nevertheless have *contingently* the property of being a physical object. Thus, he might say, there is a sense, albeit a non-literal one, in which it is true that physical objects exist contingently. For my part, however, I find this position unacceptable. It seems to me that it is just plain true – and not merely true in some special, non-literal sense – that the chair I am sitting on might not have existed.

In any case, the modal objection really concerns Quine's somewhat extreme ontological views rather than his original definition of "physical object." According to his original definition, a physical object is the material content of any region of space-time. Let's consider that proposal independently of Quine's views about the construction of matter and space-time out of the objects of pure set theory.

One immediate question we are likely to ask about this proposal is what "material" means. The problem is that it seems like "material" and "physical" ought to be synonymous, so that any definition of "physical" containing "material" is apparently circular. And even if we stipulate that "material" does not mean the same thing as "physical," so that "the material content of any region of space-time" ends up being a non-circular definition of "physical object," we will still be left in the dark about what physical objects are unless we have some independent definition of "material," and it's hard to imagine what this might be.

Still, these problems with the meaning of "material" are not fatal to Quine's proposal. We can avoid them, while at the same time retaining the spirit of his original account, by just deleting the word "material" from the original definition.[6] The result is an account of physical objects according to which a physical object is simply *the aggregate content of any portion of space-time, however ragged and discontinuous.*

This strikes me as a somewhat plausible account of physical objects, and one that yields more or less intuitive results. But I think it can be seen that the Spatial Location Account has a serious advantage over this account.

The problem with Quine's view is that it essentially amounts to a combination of the Spatial Location Account of Physical Objects with

Universalism, the answer to Peter van Inwagen's Special Composition Question according to which any non-overlapping objects whatsoever compose a single object.[7] Hence the only difference between Quine's view and the Spatial Location Account is that Quine's view entails Universalism while the Spatial Location Account does not. Nor does the Spatial Location Account entail that Universalism is false. And it seems to me that this neutrality regarding Universalism constitutes an important advantage of the Spatial Location Account over Quine's view.[8]

5. THE COMMON SENSE ACCOUNT OF PHYSICAL OBJECTS

Speaking of van Inwagen, there is a passage near the beginning of his *Material Beings* in which he suggests an account of physical objects. Van Inwagen says

> Like most interesting concepts, the concept of a material object is one without precise boundaries. A thing is a material object if it occupies space and endures through time and can move about in space (literally move about, unlike a shadow or a wave or a reflection) and has a surface and has a mass and is made of certain stuff or stuffs. Or, at any rate, to the extent that one was reluctant to say of something that it had various of these features, to that extent one would be reluctant to describe it as a material object. (Van Inwagen (1990), p. 17)

(I should note that I am assuming the expressions "physical object" and "material object" are synonymous, and also that van Inwagen takes them to be so.)

Before I continue, I need to make a short digression. As I see it, we are trying to get an analysis of the concept picked out by our usage of expressions like "physical object." But for the present purposes, there are actually two linguistic communities to consider: (i) ordinary English speakers in everyday contexts, and (ii) philosophers in discussions of philosophical issues that involve the concept of a physical object. Therefore, as I see it, there are potentially two distinct concepts involved: (i) the everyday concept of a physical object, and (ii) the philosophical concept of a physical object.

Now, I think van Inwagen's proposal is probably the correct account of the everyday concept of a physical object. I think if you ask an ordinary English speaker, or even a philosopher in an ordinary context, what physical objects are like, he or she is likely to say things that support van Inwagen's proposal. And I think this is because it just so happens that the macroscopic physical objects of our world tend to have the properties on van Inwagen's list. I will refer to the account suggested by van Inwagen, which I accept as the correct analysis of the everyday concept of a physical object, as the Common Sense Account of Physical Objects. But although I accept the Common Sense Account as the correct analysis of the everyday concept of a physical object, I think there are good reasons to prefer the Spatial Location Account as an analysis of the philosophical concept of a physical object.

One reason for preferring the Spatial Location Account to the Common Sense Account, as an account of the philosophical concept, is that, since the Common Sense Account makes the concept of a physical object an imprecise concept, it also makes any thesis about that concept – such as materialism or any of the others – an imprecise thesis. This seems to me an undesirable result. It would be better, I think, if discussions of theses like materialism turned out to be discussions of precise doctrines with determinate truth-values.

Another reason for preferring the Spatial Location Account over the Common Sense Account, as an account of the philosophical concept of a physical object, has to do with the consequences of the Common Sense Account for some of the smaller objects of our world. Van Inwagen says

> Few philosophers would be perfectly happy about calling a quark or a photon or even a large organic molecule a material object, for one has to be very careful in ascribing any of the features in the above list to such things; and talk about the surfaces of submicroscopic objects, or about the stuffs they are made of, tends to verge on nonsense. (1990, p. 17)

I don't know about other philosophers, but I for one would be perfectly happy about calling quarks, photons, and large organic molecules physical objects, despite the fact that these things lack properties on van Inwagen's list. And in fact it seems to me that the Common Sense Account's consequence that these things are not physical objects is a distinct disadvantage of that account. For these smaller objects are what objects like tables, rocks, and humans – paradigm cases of physical objects – are composed of. And surely this is a good principle concerning the mereology of physical objects (in the philosophical sense of "physical object," that is): *If x is a physical object, and y is a part of x, then y is a physical object, too.* So it seems to me that the quarks, photons and large organic molecules that compose the macroscopic, physical objects of our world must be physical objects, too, which means that the Common Sense Account simply gets the wrong result in the case of these smaller objects.

A third reason for preferring the Spatial Location Account to the Common Sense Account, as an account of the philosophical concept, is that the Common Sense Account has strange consequences concerning possible worlds with vastly different laws of nature, and very different objects, from our own world. Imagine a world in which there are objects in space, but in which the laws of nature are very different from our own laws of nature. Suppose the laws in this other world make no mention of masses, or surfaces. Suppose these laws instead talk about some alien concepts – shootspa, say, and poxie – none of which plays anything like the role played by mass (or any of the other familiar concepts) in our own laws. The laws in the alien world talk about an object's spatial location, and shootspa, and poxie, and how these factors determine what happens to that object. Imagine further that the objects in this alien world do not qualify as physical objects, according to the Common Sense Account – they have spatial locations, we can suppose, but are point-sized, pop in and out of existence instantaneously, can't move around (since they're instantaneous), don't have surfaces (being point-sized), and are not made of any stuff (again, being point-sized).

Since there would be no physical objects in this alien world, according to the Common Sense Account of Physical Objects, there would presumably be no laws of physics in the alien world, according to the proponent of the Common Sense Account. Or at any rate, if "physics" is defined as the science of physical objects and the laws of nature that govern them – and I don't know of any better definition of "physics" – then there would be no laws of physics in the alien world, since there would be no physical objects there.

But I think that, contrary to these consequences of the Common Sense Account, there would be physical objects in the alien world, at least in the philosophical sense, and I also think that the laws governing the alien objects would be the laws of physics of the alien world.

Here is some evidence for the claim that the laws governing the alien objects would be laws of physics. Imagine a team of top physicists from our world, taking advantage of the latest breakthroughs in "modal travel" technology, going to the alien world. (I guess we have to suppose, for the moment, that physicists have souls, in order to imagine them going to the alien world, since their bodies certainly couldn't exist there, on account of having masses, surfaces, etc.) Imagine our physicists making a careful study of the spatial objects in the alien world, and discovering all of the laws of nature governing those objects. It seems to me that these world-traveling physicists would be doing their thing – physics – in the alien world. And it seems to me that the laws these physicists would discover in the alien world would be laws of physics.

A defender of the Common Sense Account of Physical Objects might object that the transplanted physicists would be engaged in some other science – schmysics, say – rather than physics, and similarly that the laws they discovered would be laws of schmysics rather than laws of physics. But this goes against a very intuitive claim about the nature of physics: that the laws of physics can vary dramatically from world to world, and can even be so different in some other possible worlds that they make no mention of concepts like mass, etc.

Here is some evidence for the claim that the objects in the alien world would exemplify the philosophical concept of a physical object. Imagine a team of top philosophers from the actual world studying the alien world. (We don't really need the philosophers to be *in* the alien world, for the purposes of this thought experiment, since they can study it from the comfort of their armchairs. Thus we don't have to assume that philosophers, like physicists, have souls.) These philosophers would want to ask all the same questions that we ask about the actual world, including these: Are there any non-physical objects? Are all mental phenomena physical phenomena? Is physicalism true?

Granted, the philosophers studying the alien world may define some new concept – the concept of a schmysical object, say – and then ask whether there are any non-schmysical objects in the alien world, whether mental phenomena are schmysical phenomena in the alien world, and whether schmysicalism is true in the alien world. But I think that there are certain philosophical questions that these philosophers would want to ask concerning both the actual world and the alien world. And I think that questions about whether there are any non-physical objects, whether mental phenomena are physical phenomena, and whether physicalism is true would be among those questions.

I also think it is awkward to have to say, as the defender of the Common Sense Account has to say, that any philosophers native to the alien world, who lack the concepts of having a mass, having a surface, etc., would be unable to consider the issues that philosophers in the actual world consider when we talk about our controversies over physical objects. It seems to me preferable to be able to say that philosophers all over logical space are in a position to consider exactly the same issues about physical objects: the issue of whether there are any non-physical objects, the issue of whether all mental phenomena are physical phenomena, etc.

Here is a related reason for preferring the Spatial Location Account over the Common Sense Account as an analysis of the philosophical concept of a physical object. Suppose that the theories of the physicists are actually true, and there really are quarks and other sub-atomic particles that, as van Inwagen says, do not fit the Common Sense Account of physical objects. Does it follow from this supposition that materialism, the thesis that only physical objects exist, is false? It seems to me that it does not. And I suspect that most of the philosophers who defend materialism would be unperturbed by the discovery that the theories of the physicists are correct.

Now suppose it turns out, instead, that the true physical theory concerning the actual world posits the existence of objects without spatial locations at all.[9] And suppose that these non-spatial objects are capable of interacting with spatial objects, like our bodies. Does it follow from *these* suppositions that materialism is false? It seems to me that it does. Moreover, I think most materialists would be perturbed by the discovery that the true physical theory was such a theory. In fact, I suspect that most materialists, upon being convinced of such a discovery, would throw in the towel and concede that materialism is false, after all. And I take all of this as further evidence that the Spatial Location Account is a better analysis of the philosophical concept of a physical object than the Common Sense Account.

Here is a final reason to prefer the Spatial Location Account over the Common Sense Account as an analysis of the philosophical concept of a physical object: the Spatial Location Account allows for a way of dividing up the world's contingent objects, and characterizing the relation between the laws of nature and the laws of physics, that is very appealing for its neatness and simplicity. Let me explain.

If we adopt the Spatial Location Account then we can draw a sharp distinction between two kinds of contingent object: physical contingent objects, and non-physical contingent objects. Souls, if there were any, would be non-physical contingent objects. (I am ignoring such "derivatively contingent" objects as sets containing contingent objects, singular propositions about contingent objects, and other abstract objects that are contingent in virtue of having contingent constituents.) And we can give a precise account of the relation between two kinds of law of nature: those concerned exclusively with physical objects, and those concerned with all contingent objects, physical and non-physical. The laws concerned with all contingent objects, physical and non-physical, would be the laws of nature. And the laws concerned exclusively with physical contingent objects – the laws of physics – would be a subset of the laws of nature. This will in turn make physicalism a clear and precise thesis, rather than one concerning a concept – the concept of physics – with an indeterminate extension.

But it seems to me that adopting the Common Sense Account will not allow us to give such a tidy account of these matters. For one thing, if we accept the Common Sense Account then we cannot draw a sharp distinction between physical objects and non-physical objects, since these will be imprecise concepts with no sharp boundary separating them. And for another thing, if we accept the Common Sense Account then we will have to lump together, under the general heading of "non-physical objects," such disparate objects as quarks and electrons, on the one hand, and non-spatially-located souls, on the other hand. Similarly, if we accept the Common Sense Account then we will have no easy way of drawing a sharp distinction between the laws of nature governing physical objects and the laws of nature governing non-physical objects, since there will be no sharp boundary between these two types of object.

6. SOME OBJECTIONS TO THE SPATIAL LOCATION ACCOUNT

I'd like to conclude by considering some objections to the Spatial Location Account of Physical Objects. One objection that might be raised is this. I have given reasons to prefer the Spatial Location Account over the other accounts of the philosophical concept of a physical object, but it is not clear that I have given any reason to think that the concept of having a spatial location is in fact the concept picked out by the usage of expressions like "physical object" by actual philosophers.

My reply to this objection is that the analysis that I am proposing in this paper – the Spatial Location Account of the philosophical concept of a physical object – is, to a certain extent, meant to be a *prescriptive* rather than a *descriptive* proposal. I am not at all certain that there is in fact a single concept that all philosophers have had in mind when using the expression "physical object," but I am certain that it would be best if philosophers adopted the proposed account of what that expression means, for the reasons spelled out above. That is, I am arguing that although it might be unclear just what concept is picked out by the usage of expressions like "physical object" by actual philosophers, there is one way of resolving the muddle – namely, by adopting the Spatial Location Account – that has great advantages over its rivals. Thus it would be best for us to resolve the muddle by adopting the Spatial Location Account.

Another objection that might be raised concerns a claim that, if I am not mistaken, is actually made by a currently accepted theory in physics, namely, the claim that certain sub-atomic particles can temporarily lack any spatial location. The objection is that the Spatial Location Account would have strange consequences for such particles.

I don't find this a very troubling objection. Let's consider some different possible ways the claim in question could be understood. On one interpretation of the relevant claim, the objects in question lack *precise* spatial locations, but nevertheless have *indeterminate* spatial locations. If this is indeed the claim, then it seems to me that the Spatial Location Account would entail that these objects are indeed physical objects; and this strikes me as the right result in this case. On another interpretation of the relevant claim, the objects in question have spatial locations, but we are for some reason unable to determine what those locations are. If this is the claim, then the Spatial Location Account would again entail that these objects are physical objects; and again this strikes me as the correct result. On a third interpretation of the relevant claim, the objects in question temporarily go out of existence (or go out of existence, and are soon replaced by similar objects in nearby places). If this is the claim, then the Spatial Location Account would entail that while these objects didn't exist, they would not be physical objects; which again seems to be exactly the right result. And finally, on a fourth interpretation of the relevant claim, the objects in question really lack spatial locations at times when they exist. But if this is indeed the claim, then the Spatial Location Account entails that those objects are, at the times when they genuinely lack spatial locations, non-physical objects; and this too seems to be the correct result. It would no doubt be strange to say that some objects can change from being physical to being non-physical, and then change back again; but that just shows what a strange claim this is, when interpreted in this way.

Here is a third objection to the Spatial Location Account. There are certain entities that would count as physical objects, on the Spatial Location Account, but that we wouldn't want to count as physical objects. Among these are events, facts, spatial locations themselves, and even universals, on certain theories of universals.

I think that this is an important objection, which raises interesting issues about the entities in question, but I also think that in the end the objection is not damaging to the Spatial Location Account. Here is why. Let's grant, for the sake of argument, that entities of all of these types exist, and that some examples of each type have spatial locations. Then as a first reply to the objection, I would say that if these things deserve to be called "objects," then they are indeed physical objects, so that the Spatial Location Account gets the right result in their case. And notice that it is nice to be able to distinguish between, say, *physical* events, on the one hand, and *non-physical* events, on the other hand. Similarly with facts and universals – it is nice to be able to distinguish the physical ones from the non-physical ones.

It is also an appropriate result of the Spatial Location Account that there are no physical propositions, since propositions presumably don't have spatial locations. All of this is for the good.

But it might be thought that there is a lingering problem here that has to do with the Spatial Location Account of Physical Objects and the possibility of refuting materialism. The problem is this: if such things as propositions and the fact that 2 + 2 = 4 are *objects*, then they will count as *non-physical objects*, according to the Spatial Location Account. But their existence shouldn't be taken to show that materialists have been wrong all along.

This is indeed a problem, but if it is a problem for the Spatial Location Account then it is also a problem for the Common Sense Account, as well as any other plausible account of physical objects. After all, things such as propositions, and facts like the fact that 2 + 2 = 4, don't have mass, or most of the other properties on van Inwagen's list. Moreover, it is clear that they will count as non-physical objects on any plausible account of physical objects. Thus, on any plausible account of physical objects, these entities will serve as counter-examples to the claim that there are no non-physical objects. And yet it does not seem like these objects should refute materialism. We can call this The Problem of Non-physical Objects That Do Not Refute Materialism.

Here is my proposed solution to the problem. All parties to the dispute over the correct analysis of the concept of a physical object need some way of distinguishing between the type of object such that non-physical instances of that type present a problem for the materialist, on the one hand, and the type of object such that non-physical instances of that type do not present a problem for the materialist, on the other hand. So let that distinction be given. Then the Spatial Location Account is a way of marking the distinction between two types of object in the first category (the kind that materialists and their rivals care about). Within that category, objects with spatial locations don't refute materialism, but objects without spatial locations do.

In effect, then, I am saying that The Problem of Non-physical Objects That Do Not Refute Materialism is a problem about the formulation of materialism itself. Materialism is sometimes formulated as the view that there are no non-physical objects. But that really won't do, no matter what account of physical objects we accept. Instead, materialism should be qualified in a certain way. It should be the claim that, among "particulars," or "individuals," or "concrete objects," or whatever term we choose for characterizing the objects that materialists and their rivals care about, there are no non-physical objects. Then, given the distinction between the objects that materialists and their rivals care about and the objects that they don't care about, the different accounts of physical objects can be seen as attempts to distinguish, among the objects that materialists and their rivals care about, the ones that are acceptable to the materialists from the ones that are not.

A fourth objection to the Spatial Location Account concerns the possibility of minds with spatial locations. Such a possibility is of course ruled out by the Spatial Location Account, if we define "mind" as synonymous with "non-physical, conscious entity." But some people who accept this definition, and say they believe in minds, also believe that minds can have spatial locations. Such people will find the Spatial Location Account unacceptable.

I admit that, for those people at least, this is a disadvantage of the Spatial Location Account. But it seems to me that it is a minor disadvantage, and one that is outweighed by the various advantages of the Spatial Location Account discussed above. For the relevant people can, if they wish, simply retract the claim that the things they believe in are, strictly speaking, non-physical. Instead, they can say that they believe in conscious but physical objects with few or none of the properties on the list in the Common Sense Account of Physical Objects.[10]

NOTES

[1] Richard Cartwright quotes Hobbes in his paper "Scattered Objects," where he says "Bodies in Hobbes's sense are material objects in ours; so at any rate I shall assume." But Cartwright seems to have at least some doubts about whether Hobbes's definition is wholly correct, for he goes on to say "And I shall assume that his definition is correct at least in its implication that coincidence with some part of space is required of anything that is to count as a material object." See (1987, p. 171)

[2] For more on the alleged differences between time and space, see Markosian (1992, 1993, 1994).

[3] J. J. C. Smart (1989, p. 79) may have had something like this view in mind.

[4] It might be objected that this is a defective definition of "physics," since many other fields – biology, for example – also involve the study physical objects. But I would reply that this is no problem for the definition, since we can say that biology is the study of *living things*, some or all of which also happen to be physical. Thus all the objection shows is that certain other fields, including biology, overlap with physics.

[5] I take it that this is the definition that Berkeley would accept. He is of course notorious for going on to argue that such objects cannot exist independently of their being sensed. See his *A Treatise Concerning the Principles of Knowledge*, Part I, Sections 1ff.

[6] I'm grateful to Theodore Sider for making this point in correspondence.

[7] *Cf.* van Inwagen (1990, sections 2 and 8).

[8] It is worth noting that this version of Quine's view also entails the Doctrine of Arbitrary Undetached Parts. (For a discussion of this doctrine, see Van Inwagen (1981).

[9] This supposition is consistent with the definition of "physics" as the study of physical objects, just as the supposition that the true biological theory posits the existence of inanimate objects (oxygen molecules, for example) is consistent with the definition of "biology" as the study of living things.

[10] I am grateful to Mark Aronszajn, Ralph W. Clarke, Theodore Drange, Fred Feldman, Sharon Ryan, and Theodore Sider for help on this paper.

REFERENCES

Berkeley, G. (1982), *A Treatise Concerning the Principles of Human Knowledge* (Indianapolis: Hackett).

Cartwright, R. (1987), "Scattered Objects," in *Philosophical Essays* (Cambridge, MA: The MIT Press): 171-186.

Hobbes, T., *De Corpore.*

Markosian, N. (1994), "The 3D/4D Controversy and Non-present Objects," *Philosophical Papers* 23: 243-249.

Markosian, N. (1993), "How Fast Does Time Pass?," *Philosophy and Phenomenological Research* 53: 829-44.

Markosian, N. (1992), "On Language and the Passage of Time," *Philosophical Studies* 66: 1-26.

Quine, W. (1976), "Whither Physical Objects?," (WPO) in R.S. Cohen, P.K. Feyerabend, and M.W. Wartofsky (eds.), *Boston Studies in the Philosophy of Science* 39 (Dordrecht: D. Reidel): 497-504.

Smart, J.J.C. (1989), *Our Place in the Universe* (Oxford: Basil Blackwell).

van Inwagen, P., (1981), "The Doctrine of Arbitrary Undetached Parts," *Pacific Philosophical Quarterly* 62: 123-137.

van Inwagen, P., (1990), *Material Beings* (Ithaca: Cornell University Press).

20
TIME, SPACE AND PHYSICAL OBJECTS
A REPLY TO NED MARKOSIAN

Quentin Smith
Western Michigan University

Ned Markosian presents an interesting defense of the Spatial Location Account of physical objects. Something is a physical object if and only if it is located in space.

Markosian makes a number of good points in his paper, but I would like to raise a few questions about his account.

1. First, it seems plausible that the largest physical object is the universe, the maximal spatio-temporal whole. But the maximal spatio-temporal whole is not located in space. This is true for several reasons; the maximal spatio-temporal whole is in part temporal and time is not located in space. Second, if something is located in space, it has spatial coordinates x, y, z, but the universe has no such coordinates.

The largest physical object has the property of being spatial. But the property of being spatial is distinct from the property of being located in space. This is clear since the two questions: "where in space is the Andromeda galaxy located?" manifestly has a different meaning than "is the Andromeda galaxy spatial or non-spatial in nature?"

Markosian might respond that the universe is not a physical object. But this seems wrong and any theory that implies that the universe is not a physical object ought to be rejected in favor of a theory of physical objects that implies the universe is a physical object.

2. Shadows, mirages and the like are located in space. But they are not physical objects. I can give you the exact size, shape and spatial coordinates of the shadow of a flagpole at a certain time, but this would not imply the shadow is a physical object. If Markosian responds that a shadow is a physical object since the Spatial Location Account implies it is a physical object, I would respond that the thesis that "a shadow is not a physical object" is more plausible than any theory that implies it is a physical object.

3. Waves are located in space. Sound waves, water waves, and the like have spatial locations. But they are not physical objects but disturbances of physical objects. This is another counterexample to the Spatial Location Account.

L. N. Oaklander (ed.), The Importance of Time, 243–245.

4. Markosian writes on page 238 "Here is a third objection to the Spatial Location Account. There are certain entities that would count as physical objects, on the Spatial Location Account, but that we wouldn't want to count as physical objects. Among these are events, spatial locations themselves, and even universals, on certain theories of universals."

This seems like a good objection to me. But what is Markosian's response? He says: "Then as a first reply to the objection, I would say that if these things deserve to be called 'objects,' they are indeed physical objects, so that the Spatial Location Account gets the right result in their case" (p. 238).

I do not think this response answers the objection. The objection (in part) is that Markosian's account implies that spatially located events are physical objects, and events are not objects, physical or otherwise. Markosian's response is in effect that, if spatial events are objects, they are physical objects and his theory is immune from the objection. But this does not address the objection. The objection is that spatial events are not physical objects, and that Markosian's theory implies they are physical objects, and hence that Markosian's theory is wrong.

5. This last objection cuts to the core of one of the underlying problems with Markosian's theory. He gives us no definition of "object." He takes the meaning of "object" for granted and merely tries to define the word "physical" in the phrase "physical object." At best, this gives us a vague and ambiguous theory, inasmuch as the word "object" is vague and ambiguous. At worst, it gives a false theory, since if Markosian is using "object" to mean thing or substance, as distinct from event, property, proposition, etc., then his theory is wrong, since it implies that spatially located events are not events but things.

6. Another basic problem is that Markosian's theory seems to imply the substantival theory of space, even though he does not address this issue. If the relational theory of space is true, then his account seems to run into trouble. For a spatial location on the relational theory of space is nothing other than a physical object (not a distinct "location" at which the physical object exists). If the relational theory is true, then "a physical object has the property of being at a spatial location" means the same as the tautology "a physical object has the property of being a physical object." But then Markosian is not advancing a theory but a tautology.

Even on modal relationalism, there is a problem. Here "a physical object has the property of being at a spatial location" means "a physical object has the property of possibly being a physical object," which is a theorem of modal logic.

7. But if the substantival theory of space is true, Markosian is faced with a different set of problems. The aggregate of all spatial points is a physical

object. But this aggregate is not located anywhere; rather, all other physical objects are located in it. If the substantival theory of space is true, "being located at" is an irreflexive, intransitive and a symmetrical relation. A mass point can be located at a spatial point, but a spatial point is not located at itself, and it is not located at the mass point.

8. A further problem with Markosian's theory is that it allows that the roof of this building, Bill Clinton's nose, and the Andromeda Galaxy are together one physical object. But no such arbitrary grouping of scattered entities is a physical object. If Markosian insists it is a physical object, then he is stipulating an unorthodox sense of "physical object" and we need not be interested in it, for it does not pertain to the question, "what is a physical object?", where "physical object" has its orthodox meaning.

9. By way of conclusion, I would note that it is easy to criticize any theory of physical objects, since this is one of the hardest philosophical questions to solve. It is approximately 1,000 times harder to develop a theory of physical objects than it is to criticize someone's else theory of physical objects, and I will leave the hard work to Markosian. I do have a theory of physical objects, but I am not going to explain it, since I am not going to let Markosian give me as hard a time as I am giving him.

Part Six

TIME AND CAUSATION

21
IS PRECEDENCE A SECONDARY QUALITY?

Robin Le Poidevin
University of Leeds

A PROBLEM FOR THE TENSELESS THEORY

One of the challenges faced by anyone who proposes to revise our ordinary conception of time – and one who denies the reality of tense is surely proposing such a revision – is to show that the consequent revision can be reconciled with the facts about our ordinary experience of time (especially, in the case of the tenseless theory, those aspects of our experience which appear to point to real tense). One such fact is that our experience is temporally limited, in that what we experience, when we experience it, is always experienced as *present*. Another, related, fact is that we seem to share the same present, in that we tend to agree with each other, on the basis of our perceptions, what is going on now. Yet another fact is that we perceive *precedence*: we perceive, not just one thing that occurs after another thing we perceive, but *that* one thing occurs after another.

Tenseless theorists such as Mellor and Butterfield explain features of our temporal experience by appeal to corresponding features of causal relations. Take the three aspects identified above. We can explain one of the boundaries of our temporal experience by the fact (not undisputed) that causes are invariably earlier than their effects. Since we experience things only when they causally impinge on us, the items experienced must always be earlier than the experience itself, so we never experience the future. And since the temporal gap between an experience and the event in our immediate environment that caused it is very small (given the high velocities of light and sound), our perceptions are generally up to date – i.e. correctly represent states of affairs that obtain at the same time as those perceptions. Finally, the perception of precedence can be explained by the causal connection between successive experiences. I perceive *b* as occurring after *a*, not simply because my perception of *b* itself occurs after *a*, but because my perception of *b* is causally influenced by my perception of *a*.

Notice that these explanations appeal not simply to causality, but to causal *asymmetry*: *a*'s causing *b* rules out *b*'s causing *a*. And, as we shall

L. N. Oaklander (ed.), The Importance of Time, 249–261.

see later, there is another aspect of our experience that requires an appeal to causal asymmetry. But it is this dependence of tenseless theory on objective causal asymmetry, at least as the key to reconciling the nature of time with our experience of it, which may be that theory's undoing. For, it has been argued, real causal asymmetry requires the unreality of the future; and the unreality of the future cannot be reconciled with tenseless theory (at least as that theory is traditionally articulated). The difficulty we are now faced with has the following argumentative structure:

(1) If the tenseless theory of time is true, tense is mind-dependent.
(2) The correct explanation of (various aspects of) temporal experience requires appeal to objective causal asymmetry.
(3) The objectivity of causal asymmetry entails that the future is open.
(4) If the future is open, tense is not mind-dependent.

(1) and (4) entail:

(5) If the tenseless theory of time is true, the future is not open.

(3) and (5) entail:

(6) If the tenseless theory of time is true, then no account of temporal experience that appeals to objective causal asymmetry can be correct.

Finally, (2) and (6) entail:

(7) The tenseless theory of time is false.

Solving this difficulty for the tenseless theorist is the aim of this paper. But first, in order to show that there is indeed a difficulty to solve, I need to say something about the motivation for each of the premises.

All In the Mind: Premise (1)

The tenseless theory holds that tensed representations (e.g. thoughts, utterances, inscriptions) are made true by tenseless facts. If all representation is – ultimately – mental, then the tenseless theory entails that tense is mind-dependent. But recently, Mellor (1998) has suggested a role for tensed propositions. In his 1981, he espoused a token-reflexive treatment of tensed representations, which in the later treatment he presents as follows:

> any token of "P" is true if and only if it is as much earlier or later than *e* as "P" says the present is than *e*. (Mellor (1998), p. 31)

A "token" here would be a concrete, i.e. spatio-temporally locatable, thought, utterance or inscription. The problem that tensed theorists such as Smith (1993) have urged against this account is that it cannot accommodate the fact that at one stage it would have been true that there were no tokens of this kind, because there was at that time no mental life at all. Mellor's solution is to introduce tensed propositions, which, though they can be the contents of concrete tokens, are not dependent on there being any token which expresses them. At one time, for example, the proposition

> there are no tokens

was true. The token-reflexive account, therefore, needs to be replaced by something like the following:

> any A-[i.e. tensed] proposition "P" about any event *e* is made true at any [time] *t* by *t*'s being as much earlier or later than *e* as "P" says the present is than *e*. (1998, p. 34)

There could, then, be a tenseless semantics for tensed propositions. But insofar as such propositions could exist whether or not they were expressed, they would constitute mind-independent representations (defining a representation to be any bearer of truth or falsehood).

I am not, myself, convinced that we need introduce tensed propositions which are unexpressed by any token. In particular, I am suspicious of the inference from

> it was the case that there were no tokens

to the truth, at some time, of the proposition

> there are no tokens.

But the matter is not, in this context, of much import, since even if tense can enter into mind-independent propositions, it will not help to reconcile tenseless theory with an open future. Only if events, or concrete states of affairs, can exhibit changing tense, I want to argue, can we correctly represent the future as open, in a sense to be articulated in a later section. Since, then, Mellor's possible counterexample to (1) will not affect the argument as a whole, I suggest we ignore it in what follows.

The Failure of Causal Perspectivalism: Premise (2)

As we noted above, tenseless theorists have typically appealed to objective causal asymmetry in explaining features of our experience of time. But do they need to do so? Could causal asymmetry itself be something we project onto the world as a result of our situation and psychological constitution – could, in other words, causal asymmetry be *perspectival* in some sense? If so, then our conception of the difference between past and future, or between earlier and later, may itself be similarly perspectival. Precedence, as we might put it, could be a secondary quality. This intriguing suggestion is made by Huw Price (1997). If we can make sense of this position, then it is not necessarily a disaster for the tenseless theory that the only way of saving objective causal and temporal asymmetry is through the openness of the future, for perhaps the tenseless theory can happily wave good-bye to objective causal and temporal asymmetry.

Perspectivalism about causal asymmetry (hereafter simply "causal perspectivalism") offers to solve a nagging difficulty concerning causation, namely the failure of reductive theories of causation to account for its logical asymmetry. Here are the usual suspects:

(a) Causes are definitionally earlier than their effects, and temporal priority is logically asymmetric.

There are a number of objections to this (Humean) account of causal asymmetry. First, it rather obviously undermines the causal theory of time order – although that project is already compromised by the failure of other attempts to ground causal asymmetry. Second, commitment to the asymmetry of temporal priority automatically rules out a closed topological structure for time, but non-standard topological structures cannot be ruled out by mere definition (or can they? it depends whom one asks). Thirdly, and most seriously, it completely fails to explain the asymmetry of *dependence*: effects are dependent upon their causes, not vice versa. How could mere priority in time introduce that dependence, especially when completely unconnected events can stand in temporal relations to each other? This issue is addressed by the next account of casual asymmetry:

(b) Effects are counterfactually dependent upon causes but not vice versa

The objection here is that the account is question-begging, in that its truth depends on how we assess counterfactuals. To adapt an example of Graeme Nerlich's, suppose I notice, at 1.59 pm, that my office clock is five minutes slow. I correct it, so that at 2 pm, it is telling the right time. Intuitively, my correcting the clock at 1.59 caused it to be accurate at 2, and this is reflected in the following counterfactual: had I not corrected it at 1.59, then *in the circumstances* (i.e. given the state of the clock in the period *before* I interfered with it) it would not have been accurate at 2. The later event is counterfactually dependent upon the earlier. What, however, is wrong with the following counterfactual: had the clock not been accurate at 2, then in the circumstances (i.e. given the state of the clock in the period *after* I interfered with it) I would not have corrected it at 1.59? What is wrong is that we invariably take the relevant background circumstances, that is, those circumstances in which the cause is a necessary condition for the effect, to be those obtaining *earlier* than the effect, not those obtaining afterwards. The justification for this assumption, however, can only be in terms of the temporal direction of causation: causes occur earlier than, not later than, their effects. But we cannot simply assume that priority: it must follow from our reductive account of causation. In other words, it must be a non-trivial consequence of the counterfactual account of causal priority that causes occur earlier than their effects (if they do). Banning "backwards" counterfactuals (which make earlier events counterfactually dependent upon later ones) on the grounds that the relevant background circumstances are those that obtain earlier rather than later completely trivializes the explanation of temporal priority in terms of counterfactual priority.

(c) The laws of physics are time-reversal variant

To which the reply is, simply, that the laws of physics are mostly time-reversal *in*variant.

(d) Causes make their effects more probable; effects do not make their causes more probable.

At first sight, this is simply false. Suppose we witness an explosion within a sealed glass vessel exposed to sunlight, the result of which is a small quantity of Hydrochloric acid. We can deduce that, prior to the explosion, elemental hydrogen came into contact with elemental chlorine. The effect (explosion producing HCl) makes more probable the cause (mixing of hydrogen and chlorine in sunlight). Many other such examples could be cited.

Now there are responses to this, one of which involves appeal to the open future, and since this involves one of the other premises of the argument I will defer further discussion until a later section.

I cannot claim comprehensiveness for my brief survey of reductive accounts of causal asymmetry, but a pattern is beginning to emerge, and it is not one of resounding success. And this is part of causal perspectivalism's temptation: if, as it seems, we cannot explain causal asymmetry, perhaps we should abandon it, in favour of a view of causation as objectively symmetric, and only apparently asymmetric because of our perspective on reality.

However, causal perspectivalism faces a serious problem, and the tenseless theorist would be unwise to rely on it. An initial problem is this: unlike other secondary qualities like colour, warmth, taste and smell, we do not directly experience causal relations. This Humean point is not uncontentious (cf Anscombe), but there is in any case a reply: we *do* experience temporal relations between events, and infer causal relations from those. If the difference between earlier and later is perspectival, then so is that between cause and effect.

Here now is the real problem, which I will first state in a question-begging form (in order to show that my final statement of the problem is *not* question-begging): perspectivalism cannot explain why we perceive, and have memories of, the past, but not the future. This is question-begging because it presupposes an objective distinction between earlier and later times. Now for the *non*-question-begging form. Even if causation is intrinsically symmetric, this does not rule out causal betweenness relations obtaining. Consider, then, the following ordered (but not directed) causal series:

$$b - \beta - B$$

Perspectivalism, I submit, cannot explain why the following principle holds:

> If β is a perceptual experience, then it cannot have both b and B as its object

No mention of earlier and later here, so we are not smuggling in temporal direction. But the principle, I suggest, cannot plausibly be denied. Suppose, for example, that a perception of mine is causally intermediate between the very beginning of a sudden explosion, and my hand going up to shield my eyes. I may perceive the beginning of the explosion, or I may perceive the hand going up – but not in one and same perception. Of course, the explosion may overlap with my hand going up, but those temporal parts of the explosion which do so will not stand in the relevant relation to my perception; that is, the perception will be not be causally between those parts and my hand going up. So, what explains the truth of the principle? I can think of no other explanation than that the causal relation is intrinsically directed. If β is causally between b and B, then *either b or B* (but not both) is one of the causes of β. In other words, our experience of temporal asymmetry must have a causal explanation, and so requires causation to be objectively asymmetric.

To this it might be objected that absence of causal asymmetry need not imply absence of *any* objective asymmetry. There is, for example, a thermodynamic gradient: could this explain our experience of temporal asymmetry? I don't think it can, for the thermodynamic gradient is a global feature of the universe, not local to a particular group of causally connected states; but whatever explains the principle above, it must be intrinsic to the relations between perceptions and their objects. Causal perspectivalism, I conclude, fails, so premise (2) is safe.

Indeterminacy and the Arrow of Causation: Premise (3)

This is the most worrying premise for tenseless theorists, that we can only account for the asymmetry of causation in terms of an open, or indeterminate, future. Anyone defending this premise will, of course, need to make it clear what determinacy in this context amounts to. Here is a first attempt, in terms of bivalence: it is, at a given time *t*, determinate whether or not a certain state of affairs *s* obtains if and only if the proposition that *s* obtains is either true at *t* or false at *t*; *s* is determinate when the relevant proposition is true. (I skip over, for the moment, the question of what kind of state of affairs we are talking about here.) This we might call the *semantic* characterization of determinacy, in terms of truth and falsehood. Causal asymmetry may then consist in this: there are times at which the cause is determinate and the effect indeterminate, but no times at which the effect is determinate and the cause indeterminate. Although he talks of "fixity" and "unfixity," this seems to be the account Mackie's considers sympathetically in his (1974), the differences being that he presents it as a sufficient rather than necessary condition, and he does not characterize fixity in terms of bivalence:

> ... an effect cannot be fixed at a time when its cause is not fixed. But this has the corollary that if A and B are causally connected in a direct line, then if B is fixed at a time when A is not fixed, it must be A which is the effect and B the cause. (Mackie (1974), p. 180)

Now this might seem to present us with a perfect opportunity to explain why causal priority entails temporal priority: if the future is unfixed (indeterminate), but the past fixed (determinate), then causes will never lie in the future relative to their effects, for that would mean that there were times when effect was fixed but cause not. However, Mackie does not want to take this opportunity:

> We leave room for backward causation but not for bringing about the past: we cannot at t3 bring about A at t2, since B must already have been fixed before t3 if it is to be a cause of A. But it is conceivable that we might at t1 bring about A at t2 by way of B at t3 and the backward link from B to A: we might bring about something in the future by means of a line of causation part of which is time-reversed. (1974, p. 180)

Quite why Mackie would want to leave room for backward causation is not entirely clear, and indeed his later remarks on determinism suggest that the only circumstances in which a future event could be determinate are precisely those which rule out a direction of causation, and so rob the notion of "backward" causation of any content. The issue is taken up again below.

A rather different approach is that of Michael Tooley (1997). Tooley is explicit in his defence of the proposition that causal asymmetry requires an open future. Unlike Mackie, Tooley is not offering an *analysis* of causal asymmetry in terms of determinateness. He does, however, want to argue that causation entails certain asymmetric principles, the truth of which require what he calls a "dynamic" world. A dynamic world is one in which what facts there are depend on what time it is. Further, what facts there are at any given time does not include facts about what is the case at later times: the future, in other words, is unreal (although it is important to note that Tooley's conception of the unreality of the future is significantly different from traditional conceptions). His argument from causation to the dynamic nature of the world goes as follows:

(1) The world contains causal laws that satisfy certain probabilistic postulates relating cause and effect.

(2) These postulates are irreversible, in that substituting "cause" for "effect" and vice versa in any postulate results in a falsehood.

(3) This irreversibility cannot obtain in a static – i.e. non-dynamic – world.

Therefore: The world is dynamic.

One of these probabilistic postulates is the following:

(C_4) $\text{Prob}(Qx,\ P \Rightarrow Q\ \&\ C) = \text{Prob}(Px,\ C) + \text{Prob}(\sim Px,\ C) \times \text{Prob}(Qx,\ \sim Px\ \&\ C)$

That is: the posterior probability of the effect (i.e. the probability of the occurrence of the effect *given* the causal relationship between P and Q) in a given set of circumstances is a function of the prior probabilities, in those circumstances, of the occurrence of the cause and of the non-occurrence of the cause, together with the probability of the occurrence of the effect given the *non*-occurrence of that cause. Now (C_4) has a counterpart, (C_4*), which says that the posterior probability of the *cause* is a function of the prior probability of the *effect*, etc. The challenge is to explain why (C_4), but not (C_4*), is true. Tooley's answer is that the future is unreal, so a later state of affairs (being non-existent) cannot affect the probability of an earlier one. As he puts it, "An effect is never actual as of the time of any of its causes" (Tooley (1997), p. 113). In a static world, in contrast, where effects are actual as of the time of their causes, we have no hope of explaining the irreversibility of (C_4).

Openness Entails Tense: Premise (4)

The final premise of the argument is that talk of an open future requires appeal to mind-independent tense. The argument for this is quite simple: to describe the future as real, in contrast to the past and present, is to draw a non-relational distinction between past, present and future. It makes no sense to say that a state is real in relation to one tenseless time, but unreal in relation to another. If the future is unreal, it is so absolutely.

This connection between openness and tense, however, has been questioned. Mauro Dorato (1995), for example, proposes a formulation of the open future doctrine in which whether an object has a certain property is determinate at one tenseless time and indeterminate at earlier times. There is no appeal to tensed fact here. But consider the (putative) fact that *s* (where *s* is some type or token state) is determinate at *t*. Call that fact *f*. Now, is *f*'s determinacy similarly relative to a time? If not, then it seems we face contradiction, for if it is determinate at all times (including times earlier than *t*) that *s* is determinate at *t*, then *s* is determinate at all times; but *s* is supposed to be indeterminate at times earlier than *t*. So *f*'s determinacy must be time-relative. But the intelligibility of this is questionable. In general, the relativity of something to a context is not itself context-relative. Whether or not something is "here" is relative to one's position, but the fact that *O* is here at *l* is not position-relative. (An exception to the rule here is relativism about truth; the truth of the fact that truth is context-relative can only at best be itself context relative; but then it is not clear that relativism itself is a coherent position.) Assuming, then, that the supposed time-relativity of determinacy could not itself be time-relative, we should conclude that determinacy is not time-relative.

THE TENSELESS RESPONSE: TYPE VS TOKEN DETERMINATENESS

All four premises of the argument, then, are defensible, indeed highly plausible. So the tenseless theorist is in trouble, and it is time to consider a defensive strategy. Which premise looks the most vulnerable? I propose to attack premise (3), as I can see little room for maneuver on any other front.

Consider, then, Mackie's account of causal priority. There is, as he clearly sees, a threat to the account in the shape of determinism. Determinism, stated in nomological and time-symmetric terms, is the doctrine that, given both the total state of the universe at a given time and the laws of nature, there is for any other time *t* only one possible state of the universe at *t*. This implies that every state—past, present and future—is equally determinate:

> ....if determinism does not hold, the concept of causal priority which I have tried to analyse will apply to the objects, but if determinism holds, it will not. If you have too much causation, it destroys one of its most characteristic features. Every event is fixed from eternity with every other, and there is no room left for any preferred direction of causing. (1974, p. 191)

This is an interesting development. The kind of causation we find in Hume, for all his suspicion of causal necessity, is deterministic (admittedly in a *non*-modal sense), so that "undetermined" means "uncaused." In complete contrast Mackie is arguing that *no* causation, or at least no asymmetric causation, can be deterministic. Both positions strike me as extreme. But perhaps Mackie's position is one to which the defender of the open future is committed, for does determinism not rule out an open future? Not necessarily: not only does determinateness not entail being determined, but being determined does not entail (one kind of) determinateness.

It is important, for an understanding of the relations between determinism, determinateness, and causality, to distinguish between *type* determinateness and *token* determinateness. To put it in semantic terms, there is a distinction between the truth of a general proposition, concerning the obtaining of a type state or property, and the truth of a singular proposition, concerning a particular token of that type. Determinism entails type determinateness, but not token determinateness. The state of the universe may make it physically necessary that it will be in such-a-such a state at a later time, but how could it determine which *tokens* would realize that state? Now when we talk of the openness of the future, we could conceive this in terms of type determinateness or token determinateness: it could be indeterminate what type states will be realized in the future, or it could be indeterminate what tokens will exist in the future. If we conceive of the future as type determinate but token *in*determinate, then determinism is compatible with an open future.

Now, at first sight, this looks like good news for Mackie. Take his remark that "an effect cannot be fixed at a time when its cause is not fixed" and interpret fixity in terms of token, rather than type, determinateness. Now his account can still have application even in a deterministic universe. Or so it might seem. Certainly, if the future is open in the sense of being token indeterminate, then there will be no time at which the effect is determinate but the cause not. But this has to be made relevant to the nature of the causal relation itself. Suppose, then, that causing just is *making determinate*. It is the cause that makes determinate the effect, not vice versa. But here the remarks we made concerning determinism are relevant. Perhaps in a limited range of cases, we can make sense of something's making something else *token* determinate, but this cannot be the case in general. For example, a token coupling may make it determinate that this very token being comes into existence (since the token gametes determine the token individual). But if I leave the bath running, causing a flood, then I merely make it determinate that the type "flood" is instantiated: nothing I do determines that *this very token flood* occurs, rather than some other qualitatively indistinguishable token. We could only identify causing with making token determinate if we made the token causes part of the identity conditions of a token event, but this criterion of identity, as has been remarked before in relation to Davidson's account of event identity, is circular.

Causing, then, can at best be identified in general with making type determinate. But then, as Mackie says, we lack an explanation of causal asymmetry in a deterministic universe, for there every state, past, present or future, is as type determinate as every other. Mackie is happy to concede that there would be no direction of causation in such a universe, but I count that result as a defect of the account. If we can preserve a direction of causation even in a deterministic universe, so much the better. For one thing, it could help us to preserve a direction of time in the face of determinism.

But even if we cannot make use of token indeterminateness in our account of causation, still we might make some other use of it. Consider Tooley's probabilistic postulates. Can we appeal to the fact that there are times when the cause is token determinate and the effect token indeterminate to explain why (C_4) obtains and not (C_4^*)? This would, indeed, be one interpretation of Tooley's own explanation, that the cause is actual at times when the effect is not. I submit, however, that the issue of token determinateness is not relevant to the irreversibility of principles like (C_4). To make a, by now, familiar point: we can intelligibly ascribe a probability to a certain type state's obtaining – the probability of the volcano's erupting some time in the next five years, for example. But we cannot intelligibly ascribe a probability to a token state. (Imagine asking scientists at the scene of disaster "How likely was it that *this very token eruption* would take place?") (C_4) and its counterpart (C_4*), then, must be interpreted as principles concerning types, not tokens.

Before proceeding further, let me add a point of clarification. When I say that probability postulates concern types, not tokens, I am not denying that probability cannot apply to a single case. That is, nothing I have said, I think, prevents us from asking what, e.g., the probability is of this coin's landing tails on the very next toss. We are still asking about the probability of a certain type's being realized on this particular occasion. We are not identifying a token realization and asking what its probability is.

Now suppose that determinism is true, and that all states, at whatever time that obtain, are equally determinate. Then the fact that some states are token indeterminate will not affect the truth-values of (C_4) and (C_4*). It seems, then, that Tooley's explanation of the irreversibility of the probabilistic postulates requires that the universe be indeterministic. As with Mackie's account, I think this is an unfortunate limitation.

Let us assume, however, that the world is indeed indeterministic, and, moreover, that the past is type determinate (by virtue of being token determinate), and the future type indeterminate (by virtue of being neither determined nor token determinate). Is the irreversibility of (C_4) and similar postulates thereby explained? I don't think it is. For even if it is, at the time of the cause, indeterminate whether or not a certain (type) effect will occur, the *probability* of that effect is not. This assumes, admittedly, that indeterminism is compatible with the *chances* of things being determined, but I am unapologetic about that assumption, as it seems to me that (C_4) requires that chances are indeed determined. Now, if the chances of an effect are determinate (because determined), then I cannot see why the fact that those chances are less than 1 at times when the cause is actual should, by itself, prevent us from saying that the posterior probability of the cause is a function of the prior probability of the effect.

Of course, there is a very good reason why we think it is the cause which determines the probability of the effect and not vice versa, and that is simply because the causal relation runs only one way. It is the asymmetry of causation that makes the relevant probability postulates irreversible. Type and token indeterminacy are simply beside the point. But if that is so, then it seems that the asymmetry of causation (or asymmetric aspects of causation) do not require us to view the future as open. The tenseless theory is safe – for the time being, at any rate.

REFERENCES

Butterfield (1984), "Seeing the Present," *Mind* 93: 161-76; reprinted with corrections in R. Le Poidevin (ed.), *Questions of Time and Tense* (Oxford: Clarendon Press): 61-75.

Dorato, M. (1995), *Time and Reality: Spacetime Physics and the Objectivity of Becoming* (Bologna: CLUEB).

Mackie, J. L. (1974), *The Cement of the Universe: A Study of Causation* (Oxford: Clarendon Press).

Mellor, D. H. (1998), *Real Time II* (London: Routledge).

Price, H. (1997), *Time's Arrow and Archimedes' Point* (Oxford: Clarendon Press).

Tooley, M. (1997), *Time, Tense, and Causation* (Oxford: Clarendon Press).

22

IS PRECEDENCE A SECONDARY QUALITY? A REPLY TO ROBIN LE POIDEVIN

Quentin Smith
Western Michigan University

1. WHY CAUSAL ASYMMETRY MAY BE A PROBLEM FOR THE TENSELESS THEORY OF TIME

Le Poidevin's argument is that defenders of the tenseless theory of time must explain the aspects of our experience of time that seem to suggest that the tensed theory of time is true. In particular, he mentions (1) what is experienced is always experienced at present; (2) we seem to share the same present and (3) we perceive that one thing occurs after another. Regarding the third point, we perceive something y as occurring after something x because in part our perception of y is causally influenced by our perception of x.

Tenseless theorists, such as Mellor and Butterfield, explain these facts of our experience in terms of causal asymmetry. Robin Le Poidevin raises an interesting and novel question, namely, that this explanatory project may raise a problem for the tenseless theorist. The problem is that causal asymmetry may require the unreality of the future, in which case the tenseless theory of time is false.

In particular, it could be that:

(1) causality is asymmetric and
(2) asymmetric causality implies the future is open and
(3) the openness of the future implies tense is mind-independent.

Le Poidevin presents a somewhat different argument at the beginning of his paper, but I think the tenser would be happy with holding merely that the tense is not mind-dependent. I would be content to rest with (1) - (3) as one kind of sound argument for the truth of the A-theory or tensed theory of time. Le Poidevin plausibly argues that causality is asymmetric and he analyses very carefully the idea of whether this implies the future is open. The statement that "the future is open" can be understood in terms of indeterminateness (not to be confused with causal indeterminism). It is, at

L. N. Oaklander (ed.), The Importance of Time, 263–266.

time t, indeterminate whether or not a certain state of affairs s obtains if and only if it the proposition that s obtains is neither true nor false at t.

I think Le Poidevin plausibly argues, contra Mauro Dorato (1995) that there are tensed facts if the future is open. Consider the tenseless fact F that a state of affairs s is tenselessly determinate at t. The fact F is determinate at all times, and thus s is determinate at all times, which would preclude s from being indeterminate at times earlier than t. Only if it is a transiently present tensed truth that s is determinate, and a transiently past tensed truth that s was indeterminate, can the future be said to be open at some time earlier than the time at which s is determinate.

2. IS CAUSAL ASYMMETRY PRIMITIVE?

The proposition that Le Poidevin proposes to criticize, in order to save the tenseless theory of time, is the proposition that (3) "The objectivity of causal asymmetry entails that the future is open."

I would respond at first that I think that the objectivity of causal asymmetry does not entail the future is open. Consider that the propositions describing what we experience or observe cannot all be infinitely analyzable, for if they were, we could not understand any of them, for we lack the time to grasp the infinite number of concepts involved in the analysis.

Furthermore, there must be some propositions characterizing our experience that we grasp; if there were not, it would be true that our experience is correctly describable as indescribable, which appears to be a self-contradiction.

Given, this some propositions describing our experience are finitely analyzable. This implies they contain some primitive concepts, or primitive properties or relations. I think one primitive relation is causation. When I push a bottle, I observe that my pushing effort is causing the bottle to move. Further, I observe that my pushing effort is asymmetric; the causal relation between my pushing and the moving bottle is such that the pushing causes the movement of the bottle, but the movement of the bottle does not cause the pushing.

So, I think it is a primitive fact of experience that the causal relation is asymmetric. Whether or not this experience includes an experience of an open future is a matter for debate. I think the defender of the tenseless theory of time could say that the tenser and detenser could both agree it is a fact of immediate experience that the causal relation is asymmetrical, and make this agreement independent of any further claims about alleged observations that the future is open or closed.

I shall not make a further claim on behalf of the tenser here, for otherwise I can simply defend the tensed theory of time by saying that I observe it to be true.

Indeed, I shall not even rest my case about causal asymmetry upon observation, since it is very easy to assert that the justification for my belief that causation is asymmetrical is that "I observe it to be asymmetrical." It would also be very easy for someone to deny that they observe this and we are then left with little more than an exchange of what each of us seems to introspect.

I think Le Poidevin's path is better, where some argument is given against the thesis that "The objectivity of causal asymmetry entails that the future is open."

3. CAUSAL INDETERMINISM, INDETERMINATENESS AND PROBABILITY

Le Poidevin's argument is that if the world is causally indeterministic, then it still could be that the probability of the effect is never indeterminate. This is the case even if at the time of the cause, it is indeterminate whether or not a certain (type) effect will occur. I assume this is what Le Poidevin means when he writes: "For even if it is, at the time of the cause, indeterminate whether or not a certain (type) effect will occur, the probability of that effect is not." I take this to mean:

At time t, it is indeterminate to degree n whether or not a certain type effect will occur, but it is determinate to degree one that it is indeterminate to degree n whether or not that type effect will occur.

I think that is what Le Poidevin is getting at, and I see no problem with that statement. Indeed, it seems to be implied by standard interpretations of quantum mechanics.

But following this there are two interesting statements. (1) Le Poidevin says that it is the asymmetry of causation that makes the relevant probability postulates irreversible. (2) Type and token indeterminacy are beside the point, and we are not required to view the future as open.

What I have not grasped is what these statements are derived from. What are their premises? I cannot detect them in the preceding discussion of type and token indeterminacy or determinacy regarding causes, effects and probabilities. It seems almost as if it is a mere assertion that [1] "the asymmetry of causation makes the relevant probability postulates irreversible").

But, although I think Le Poidevin may well be right that the asymmetry of causation is what makes the probability postulates irreversible, my only reason for thinking this is that I immediately experience causation to be asymmetrical, which determines irreversibility. I cannot see what Le Poidevin's reasons are for thinking this. If pressed, I would think his argument would that be that he has previously argued that (a) causality is asymmetrical, that (b) there must be some explanation of the irreversibility of the probability postulates, and that (c) the best explanation is that causality is asymmetrical. But I am not sure if this is exactly the structure of his argument.

4. CAUSATION OF TOKENS

There is another interesting distinction that Le Poidevin makes in the course of reaching this conclusion, namely, between the type and token openness of the future. This distinction seems plausible at first glance, namely, that the future can be type determined but not completely token determined. Maybe this is true. Let us see if we can think of a counterexample.

Suppose there are three interrelated electrons that constitute the entire initial conditions of the universe. These are token-particulars. Given the law of the conservation of matter, it is these same three electrons that persist throughout the entire future of the universe. Given that the laws of nature are causally deterministic, it would follow that: For every time t1 at which the cause is token determinate, the effect at a later time t2 is also token determinate, inasmuch as the effect consists in nothing other than these same three electron-tokens standing in some relation to each other.

Put in semantic terms, why cannot we have as one premise a causally deterministic law of nature, a second premise a singular, Kaplan-like proposition that includes the three electrons and the relation among them at time t1. This would be a deductive-nomological argument in Hempel's sense. We then deduce from these general and singular premises a singular conclusion, namely, a conclusion that consists of the singular proposition that includes all three electrons as related in a way specified by the two premises. Would not this be a case where the future t2 is token determinate, as well as type determinate? And if not, why not?

We cannot say that there could be different electrons, for that would violate the law of matter conservation, which implies for elementary particles (which have no parts) that they cannot be destroyed; it is the same elementary particles that persist through time, not qualitatively similar ones (which would require that at some time some electrons be destroyed and others created, in violation of the law of conservation). In this universe, the entire history of the universe would be completely both causally determinate and token determinate.

If I am right, I don't thing this would affect Le Poidevin's basic argument, for at the end he concludes that "type and token indeterminacy are simply beside the point," all that we need, Le Poidevin seems to be saying, is the thesis that causal asymmetry explains why the probability postulates are irreversible. And I don't see any reason to think that this conclusion is implausible.

23
RESPONSE TO ROBIN LE POIDEVIN'S PAPER, IS PRECEDENCE A SECONDARY QUALITY?

Michael Tooley
University of Colorado at Boulder

1. A PROBLEM FOR THE TENSELESS THEORY

The argument on which Le Poidevin focuses in his paper is as follows:

(1) If the tenseless theory of time is true, tense is mind-dependent.
(2) The correct explanation of (various aspects of) temporal experience requires appeal to objective causal asymmetry.
(3) The objectivity of causal asymmetry entails that the future is open.
(4) If the future is open, tense is not mind-dependent.

(1)and (4) entail:

(5) If the tenseless theory of time is true, the future is not open.

(2)and (5) entail:

(6) If the tenseless theory of time is true, then no account of temporal experience that appeals to objective causal asymmetry can be correct.

Finally, (2) and (6) entail:

(7) The tenseless theory of time is false.

2. LE POIDEVIN'S DISCUSSION OF PREMISE (1): THE CONNECTIONS BETWEEN TENSELESS VIEWS AND THE MIND-DEPENDENCE OF TENSE

Le Poidevin points out that "... recently, Mellor (1998) has suggested a role for tensed propositions." He goes on to say, however, that we need to

L. N. Oaklander (ed.), The Importance of Time, 267–284.

introduce tensed propositions, and on this matter I agree with Le Poidevin. For, in the first place, all that one needs to do is to provide truth conditions for tensed utterances, thoughts, beliefs, etc., and this can be done in a way that escapes Quentin Smith's objection. For consider the sentence:

> "Once there were no tokens."

If this is analyzed along the lines of

> Once there was a time when there was a true token of the sentence "There are now no tokens"

the account is obviously unsatisfactory. But one can avoid this problem if one adopts, instead, the following analysis:

> "There exists (tenselessly) a time t such that time t is earlier than the present, and there are no tokens at time t."

Secondly, the postulation of propositions – tensed or otherwise – is a controversial matter, and the situation would not be a happy one if one could avoid Quentin Smith's objection only by postulating that existence of propositions.

Notice, finally, that Mellor runs into problems because he models his account of the truth conditions of past tensed sentences on Prior's account of the logical syntax of such sentences – according to which it is a matter of applying a past tense operator to a core sentence that is in the present tense, and that involves an indexical.

3. LE POIDEVIN'S DISCUSSION OF PREMISE (2): THE FAILURE OF CAUSAL PERSPECTIVALISM

Le Poidevin considers the following four, reductionist accounts of causation, and argues that all of them are problematic:

(1) Causes are definitionally earlier than their effects, and temporal priority is logically asymmetric.

(2) Effects are counterfactually dependent upon causes but not vice versa.

(3) The laws of physics are time-reversal variant

(4) Causes make their effects more probable; effects do not make their causes more probable.

Le Poidevin's objection to the second reductionist view – i.e., to an analysis based upon the idea that effects are counterfactually dependent upon causes but not vice versa – overlooks the possibility of a Stalnaker/Lewis style analysis of counterfactuals. But a Stalnaker/Lewis style analyses of counterfactuals are, in fact, open to decisive objections, so Le Poidevin's discussion could easily be supplemented at this point.

Le Poidevin, in discussing a counterfactual analysis of causation, says that "it must be a non-trivial consequence of the counterfactual account of causal priority that causes occur earlier than their effects (if they do)." But if one considers a world where one supposes that backward causation is possible, then one will have true counterfactuals of the form:

> "If C had occurred at time t_2 then E would have occurred at the earlier time t_1."

In discussing the idea of a probabilistic approach to causation, Le Poidevin argues that if causation is to be asymmetric, it will not suffice that causes make their effects more probable; it will also have to be the case that effects do not make their causes more probable, and he maintains that many counterexamples to the latter claim can be advanced.

The problem here, however, is more general than Le Poidevin indicates. For suppose that a certain sort of cause – C – makes its effect – E – more probable, so that one has

$$\text{Prob}(E/C) > \text{Prob}(E)$$

In view of the definition of conditional probabilities, one has:

$$\text{Prob}(E/C) \times \text{Prob}(C) = \text{Prob}(C \,\&\, E) = \text{Prob}(C/E) \times \text{Prob}(E)$$

Therefore

$$\frac{\text{Prob}(E/C)}{\text{Prob}(E)} = \frac{\text{Prob}(C/E)}{\text{Prob}(C)}$$

From which it follows that

$$\text{Prob}(E/C) > \text{Prob}(E) \text{ if and only if } \text{Prob}(C/E) > \text{Prob}(C).$$

So causes raise the probabilities of their effects if and only if effects raise the probabilities of their causes.

Given the problems that reductionist approaches to causation face – and there are others besides those that Le Poidevin mentions, including objections based upon *simple* universes, and *temporally inverted* universes – the question naturally arises as to why one should not abandon reductionist accounts in favor a realist account.

Le Poidevin contends that "we *do* experience temporal relations between events…." I am not sure whether he claiming here that we *directly* experience temporal relations between events But if he is, then I think that a persuasive argument can be mounted against that view, and in favor of the view that beliefs about events being temporally ordered involve a combination of perception and (very short term) memories.

(Le Poidevin next offers an objection to causal perspectivalism – where this is the view that "causation is objectively symmetric, and only apparently asymmetric because of our perspective on reality." His objection involves supposing that there are at least relations of causal betweenness, and then considering the following ordered (but not directed) causal series –

b–β–B

and Le Poidevin contends that perspectivalism cannot explain why the following principle holds:

> If β is a perceptual experience, then it cannot have both b and B as its object.

But what is meant by the *object* of a perceptual experience? If the term "object" means something that is part of the cause of a perceptual experience, that the advocate of causal perspectivalism could say that the principle

> If β is a perceptual experience, then it cannot have both b and B as its object.

is trivially true, since a perceptual experience cannot have *either* b or B as its object. But perhaps the idea of an object of an experience can be understood in terms purely of a *nomological mapping* relation between an external object and an internal image.

I would prefer to argue against causal perspectivalism in two slightly different ways. First of all, consider complex nomological forks, such as those that arise when a stone strikes the surface of a pond, or where a light is turned on. Why is that all such complex nomological forks are *open towards the future*? Causal perspectivalism, I believe, can provide no satisfactory answer.

Secondly, if there is no direction to causation, then such complex nomological forks appear to be *extremely improbable*, whereas if there is a direction to causation, and if causation satisfies certain postulates relating the probability of an effect to the probability of its cause, then one can show that such complex nomological forks are not highly improbable.

4. LE POIDEVIN'S DISCUSSION OF PREMISE (3): INDETERMINACY AND THE ARROW OF CAUSATION

Le Poidevin offers the following, very concise statement of my own argument:

(1) The world contains causal laws that satisfy certain probabilistic postulates relating cause and effect.
(2) These postulates are irreversible, in that substituting "cause" for "effect" and vice versa in any postulate results in a falsehood.
(3) This irreversibility cannot obtain in a static – i.e. non-dynamic – world.

Therefore: The world is dynamic.

The argument then focuses upon one of these probabilistic postulates, namely:

(C_4) $\text{Prob}(Qx, P \Rightarrow Q \;\&\; C) = \text{Prob}(Px, C) + \text{Prob}(\sim Px, C) \times \text{Prob}(Qx, \sim Px \;\&\; C)$

But "(C4) has a counterpart, (C4*), which says that the posterior probability of the cause is a function of the prior probability of the effect, etc. The challenge is to explain why (C4), but not (C4*), is true."

"Tooley's answer is that the future is unreal, so a later state of affairs (being non-existent) cannot affect the probability of an earlier one. As he puts it, "An effect is never actual as of the time of any of its causes" (Tooley (1997), p. 113). In a static world, in contrast, where effects are actual as of the time of their causes, we have no hope of explaining the irreversibility of (C4).

5. LE POIDEVIN'S DISCUSSION OF PREMISE (4): OPENNESS ENTAILS TENSE

5.1 Le Poidevin's Argument

Le Poidevin offers the following argument for premise (4):

> The final premise of the argument is that talk of an open future requires appeal to mind-independent tense. The argument for this is quite simple: to describe the future as real, in contrast to the past and present, is to draw a non-relational distinction between past, present and future. It makes no sense to say that a state is real in relation to one tenseless time, but unreal in relation to another. If the future is unreal, it is so absolutely. (p. 257)

To get clear about this argument, it is essential to get clear about what is meant by the crucial sentence:

> It makes no sense to say that a state is real in relation to one tenseless time, but unreal in relation to another. (p. 257)

An advocate of a tensed view of time is committed to a dynamic view of reality, according to which the states of affairs that are actual as of one time may not be actual as of another. But Le Poidevin refers here to *tenseless* times, and so the present claim seems to come to this:

> If time is tenseless, then it makes no sense to say that a state of affairs is real as of one time, but unreal as of another.

Le Poidevin mentions that Mauro Dorato (1995), has proposed a formulation of the open future doctrine in which whether an object has a certain property is determinate at one tenseless time and indeterminate at earlier times. Le Poidevin argues, however, that this is problematic:

> But consider the (putative) fact that *s* (where *s* is some type or token state) is determinate at *t*. Call that fact *f*. Now, is *f*'s determinacy similarly relative to a time? If not, then it seems we face contradiction, for if it is determinate at all times (including times earlier than t) that *s* is determinate at t, then *s* is determinate at all times; but *s* is supposed to be indeterminate at times earlier than t. So *f*'s determinacy must be time-relative. But the intelligibility of this is questionable. In general, the relativity of something to a context is not itself context-relative. Whether or not something is "here" is relative to one's position, but the fact that *O* is here at *l* is not position-relative. (An exception to the rule here is relativism about truth; the truth of the fact that truth is context-relative can only at best be itself context relative; but then it is not clear that relativism itself is a coherent position.) Assuming, then, that the supposed time-relativity of determinacy could not itself be time-relative, we should conclude that determinacy is not time-relative. (p. 257)

5.2 Comments on Le Poidevin's Argument

How are we to understand the idea of being determinate at a time? It appears that what Le Poidevin has in mind here might be explained, either in terms of the notion of being actual as of a time, or in terms of the notion of being true at a time, along something like the following lines:

"State of affairs *s* is determinate at time *t*"

means the same as

"*s* is part of what is actual as of time *t*"

"State of affairs *s* is determinate at time *t*"

means the same as

"It is true at time *t* that *s* exists"

Suppose now that *s* is an instantaneous state of affairs that exists *at* time *t1*, and let *t0* be any earlier time. Then state of affairs *s* will, on the view being considered here, be determinate at time *t1*, but not at time *t0*. If we express this in terms of the idea of truth at a time, what we have is this:

(1) "It is true at time *t1* that *s* exists"
(2) "It is not true at time *t0* that *s* exists"

What, now, is the status of *f* – where *f* is *s*'s being determinate at time *t1*? For *f* to be determinate at the times *t1* and *t0* will be for the following to be the case:

(3) "It is true at time *t1* that it is true at time *t1* that *s* exists"
(4) "It is true at time *t0* that it is true at time *t1* that *s* exists"

But given that *s* is an instantaneous state of affairs that exists *at* time *t1*, the only thing that can make it true at time *t1* that *s* exists is something – namely *s* – that exists at time *t1*. This, however, will not be part of what is actual as of the earlier time *t0*, and so that is actual as of time *t0* cannot contain any truthmaker for the proposition that it is true at time *t1* that *s* exists. So we have the negation of (4):

~(4) "It is *not* true at time *t0* that it is true at time *t1* that *s* exists"

The upshot is that *f* is not determinate at times earlier than *t1*. But this is not mysterious, since it follows, as we have just seen, from the concept of being true at a time, together with the metaphysical claim that the future is not real.

6. THE TENSELESS REPONSE: TYPE VERSUS TOKEN DETERMINATENESS

Le Poidevin is right, I believe, in thinking that any attack upon the argument that he has set out must be directed against premise (3).

Le Poidevin examines Mackie's account of causation in terms of "fixity," and he points out that it entails that if the world is deterministic, then no events are causally related. Le Poidevin is right in thinking that this outcome is not acceptable.

Le Poidevin is also right in thinking that the following thesis is true:

(a) Determinateness not entail being determined.

But what about the converse thesis:

(b) Being determined does not entail determinateness?

I believe that that thesis is also true. Le Poidevin, however, wants to advance a restricted version of this second thesis, namely:

(c) Being determined does not entail (one kind of) determinateness.)

His reason for doing so is as follows:

> It is important, for an understanding of the relations between determinism, determinateness, and causality, to distinguish between type determinateness and token determinateness. To put it in semantic terms, there is a distinction between the truth of a general proposition, concerning the obtaining of a type state or property, and the truth of a singular proposition, concerning a particular token of that type. Determinism entails type determinateness, but not token determinateness. The state of the universe may make it physically necessary that it will be in such-and-such a state at a later time, but how could it determine which *tokens* would realize that state? (p. 258)

It seems to me that there is a strong objection to this view that determinism entails type determinateness, but not token determinateness, and which is as follows. First, suppose that there are conservation laws that entail that the existence of an electron at time t_1 will causally give rise to there being an electron throughout the interval (t_1, t_2) that is open at the one end and closed at the other. What is the relation between the electron that exists at time t_1 and the electron that exists at time t_2? If identity over time is logically supervenient upon relevant causal relations – and it surely is – then the electron that exists at time t_1 and the electron that exists at time t_2

will be one and the same electron. But then *if* determinism in this case entails type determinateness, must it not entail token determinateness as well, given that the relevant causal relations fix the identity of the second electron? Secondly, is the situation different if one considers a type of thing that exists at time t_2, but not at time t_1? I cannot see that it is, since it seems to me that the identity of that new thing will be fully fixed by its causal relations to the states of affairs that were causally sufficient to bring it into existence. So, once again, *if* determinism in this case entails type determinateness, it seems to me that it must entail token determinateness as well.

I am disagreeing, then, with what Le Poidevin says in the following passage:

> But if I leave the bath running, causing a flood, then I merely make it determinate that the type "flood" is instantiated: nothing I do determines that *this very token flood* occurs, rather than some other qualitatively indistinguishable token. We could only identify causing with making token determinate if we made the token causes part of the identity conditions of a token event, but this criterion of identity, as has been remarked before in relation to Davidson's account of event identity, is circular. (p. 259)

My response to this counterargument is that what is needed to support the claim the very same token flood occurs is not a general analysis of identity that covers *both* identity at a time *and* identity over time, but only an analysis of identity over time. For that enables one to say that given that some (possibly very complex) state of affairs C existing at time t_1 is a totally sufficient cause of some state of affairs E existing at time t_2, if the world had been different in some way that did not affect state of affairs C existing at time t_1, then the world would still have contained state of affairs E, and not merely some state of affairs qualitatively indistinguishable from E.

The upshot of this is that it is not true that causation "can at best be identified in general with making type determinate." One might, instead, try identifying it with making *token* determinate.

But let me immediately add that I do not think that either route is really promising. For what is meant by the term "making?" In particular, can it be anything other than a synonym for *causing*?

The identity over time point also bears upon the following passage:

> But even if we cannot make use of token indeterminateness in our account of causation, still we might make some other use of it. Consider Tooley's probabilistic postulates. Can we appeal to the fact that there are times when the cause is token determinate and the effect token indeterminate to explain why (C4) obtains and not (C4*)? This would, indeed, be one interpretation of Tooley's own explanation, that the cause is actual at times when the effect is not. I submit, however, that the issue of token determinateness is not relevant

> to the irreversibility of principles like (C4). To make a, by now, familiar point: we can intelligibly ascribe a probability to a certain type of state obtaining (the probability of the volcano's erupting some time in the next five years, for example. But we cannot intelligibly ascribe a probability to a token state. (Imagine asking scientists at the scene of disaster "How likely was it that this very token eruption would take place?") (C4) and its counterpart (C4*), then, must be interpreted as principles concerning types, not tokens. (pp. 259-260)

For imagine a case involving identical twins at a very early stage in development, where, at time *t*, one of the sets of cells is destroyed, and the other survives. Consider, then, a slightly different scenario, where exactly one twin will be destroyed, but where each has a 50:50 chance of being destroyed. Then it is true, it seems to me that the probability of there being an individual of the same type as the individual who actually survived is 100%, while the probability of the very same individual surviving is 50%. So one can refer to the probability of a token event, as well as to the probability of an event of a given type.

But while disagreeing with Le Poidevin's claim about the applicability of probability to a token event, I do want to agree with the conclusion that the distinction between types and tokens is not crucial with respect to postulate (C4) and its counterpart (C4*).

7. THE CRUCIAL ISSUE: THE ARGUMENT BASED ON THE CAUSAL POSTULATES

7.1 The Argument from Causation for a Tensed/Dynamic World

In the initial draft of my book, *Time, Tense, and Causation*, I attempted to argue for a dynamic or tensed approach to time according to which the past and the present are real, but the future is not by appealing to the idea that something is not a fact at a given time if it would have been logically possible for there to be someone at that time with the power to prevent the state of affairs in question from obtaining. I became convinced, however, that this line of argument – which I referred to as the Argument from Preventability – was open to two strong objections. One objection was, first, that this line of argument could only succeed if one could offer a proof of the impossibility of backward causation – and one, moreover, that did not presuppose a tensed view of time – and, secondly, that no metaphysically neutral proofs of the impossibility of backward causation that have so far been offered are successful. (The reason is that even if such proofs are successful, they seem to establish at most that causal loops are impossible, and one can have cases of backward causation in worlds where backward causation is nomologically impossible.) A second objection was that an advocate of a tenseless view of time can provide truth conditions for the crucial counterfactuals that enter into the Argument from Preventability, and so it cannot be true that the idea of a dynamic world can be captured by those counterfactuals.

These two objections to the Argument from Preventability suggested, however, two other lines of argument – one of which focused upon the truth conditions of counterfactuals, and the other of which centered upon causation. The latter argument – which is the one that Le Poidevin has considered – rested upon a realist account of causation, which can be formulated as follows:

Causation is that unique relation between events or states of affairs such that if there are laws involving that relation, those laws must satisfy the following four postulates:

(C_1) $\text{Prob}(Px, P \Rightarrow Q \ \& \ C) = \text{Prob}(Px, C)$

(C_2) $\text{Prob}(Qx, P \Rightarrow Q \ \& \ C) = \text{Prob}(Px, C) + \text{Prob}(\sim Px, C) \times \text{Prob}(Qx, \sim Px \ \& \ P \Rightarrow Q \ \& \ C)$

(C_3) $\text{Prob}(Qx, \sim Px \ \& \ P \Rightarrow Q \ \& \ C) = \text{Prob}(Qx, \sim Px \ \& \ C)$

(C_4) $\text{Prob}(Qx, P \Rightarrow Q \ \& \ C) = \text{Prob}(Px, C) + \text{Prob}(\sim Px, C) \times \text{Prob}(Qx, \sim Px \ \& \ C)$

Given this account of causation, the Argument from Causation can be formulated as follows:

> First, our world contains many types of events that prima facie, and in the absence of a certain type of explanation, are *extremely* improbable. Secondly, although the events in question do have causal explanations, that fact entails that they are not, all things considered, extremely improbable events, if and only if causal laws satisfy postulates (C_4) through (C_4). So, thirdly, it is reasonable to believe that our world does contain causal laws that satisfy those postulates. But, fourthly, it is possible to explain how causal laws can satisfy those postulates if, and only if, the world is dynamic rather than static. Therefore, it is reasonable to believe that our world is a dynamic one, where what facts there are depends upon what time it is.

7.2 Le Poidevin's Formulation of the Problem, and his Objection to the Argument from Causation

The crucial step in the argument is the fourth one, where I claim that it is possible to explain how causal laws can satisfy postulates (C_1) through (C_4)

if, and only if, the world is dynamic rather than static. In setting out the argument for that claim, I indicate that an argument that shows how postulate (C_4) can be true must be such that it also justifies the following postulate

$$(C_4^*) \quad \text{Prob}(\sim Px,\ P \Rightarrow Q\ \&\ C) = \text{Prob}(\sim Qx,\ C) + \text{Prob}(Qx,\ C) \times \text{Prob}(\sim Px,\ Qx\ \&\ C)$$

since the latter postulate is incompatible with postulate (C_1). I then go on to argue that if the world were a dynamic one where the past and the present are real, but the future not, then postulates (C_1) through (C_4) would be true, but not (C_4*).

Le Poidevin tries to show that the argument that I offer is problematic, and here he offers two considerations. The first is this:

> Now suppose that determinism is true, and that all states, at whatever time that obtain, are equally determinate. Then the fact that some states are token indeterminate will not affect the truth-values of (C4) and (C4*). It seems, then, that Tooley's explanation of the irreversibility of the probabilistic postulates requires that the universe be indeterministic. As with Mackie's account, I think this is an unfortunate limitation. (p. 260)

Le Poidevin is absolutely right that such a limitation would be very unfortunate. But my account of causation does *not* entail that there is no causation in a deterministic world. The reason is that Le Poidevin's argument turns upon the notion that determinism entails that all events are type determinate. The notion that my argument involves, however, is not the notion of an event being type determinate – where the latter is so understood that it is entailed by determinism – but the idea of an event or state of affairs being actual as of a time, and as I understand that notion, the existence of a state of affairs C that exists at time t_1 and that is causally sufficient for a state of affairs E at a later time t_2 does not entail that state of affairs E is actual as of the earlier time t_1. Because of this, my analysis of causation is, unlike Mackie's, compatible with determinism.

Le Poidevin's second argument is contained in the following passage"

> Let us assume, however, that the world is indeed indeterministic, and, moreover, that the past is type determinate (by virtue of being token determinate), and the future type indeterminate (by virtue of being neither determined nor token determinate). Is the irreversibility of (C4) and similar postulates thereby explained? I don't think it is. For even if it is, at the time of the cause, indeterminate whether or not a certain (type) effect will occur, the probability of that effect is not. This assumes, admittedly, that indeterminism is compatible with the chances of things being determined, but I am unapologetic about that assumption, as it seems to me that (C4) requires that chances are indeed determined. Now, if the chances of an effect are

> determinate (because determined), then I cannot see why the fact that those chances are less than 1 at times when the cause is actual should, by itself, prevent us from saying that the posterior probability of the cause is a function of the prior probability of the effect. (p.260)

At the heart of this argument is the idea that "even if it is, at the time of the cause, indeterminate whether or not a certain (type) effect will occur, the probability of that effect is not," and when Le Poidevin goes on to claim that the posterior probability of the cause may very well be a function of the prior probability of the effect, what he is calling "the prior probability of the effect" is the probability of the effect, *given the occurrence of the cause.* But such probabilities cannot be used to undermine the argument that rests upon postulates (C_1) through (C_4), since those postulates involve a crucial restriction upon the additional evidence that is represented by C:

> Finally, I need to refer to information of a certain restricted sort – specifically, information that either is tautological, or that concerns only what causal laws there are. I shall use the term C for that purpose. (Tooley (1997) p. 59)

This restriction on postulates (C_1) through (C_4) is essential, since otherwise it is easy to generate contradictions. Given this restriction, one cannot make use of the probability of an effect, given the occurrence of its cause, to show that – contrary to postulate (C_1) – the posterior probability of a cause is a function of the prior probability of its effect.

7.2 Le Poidevin's Alternative Proposed Solution to the Problem

Le Poidevin also has, however, another important objection to my argument. The thrust of it is that one can explain why it is that (C_4) is true while (C_4*) is not *without* appealing to the idea of a dynamic world. How can this be done? Le Poidevin's answer is given in the final paragraph of his paper:

> Of course, there is a very good reason why we think it is the cause which determines the probability of the effect and not vice versa, and that is simply because the causal relation runs only one way. It is the asymmetry of causation that makes the relevant probability postulates irreversible. Type and token indeterminacy are simply beside the point. But if that is so, then it seems that the asymmetry of causation (or asymmetric aspects of causation) do not require us to view the future as open. The tenseless theory is safe–for the time being, at any rate. (p. 260)

This proposal is open, however, to two possible objections. The first concerns how one is to establish that causation is asymmetric, and whether that can be done in a way that is compatible with a tenseless approach to time. What is, I think, clear, is that one cannot do this by establishing the more general thesis that backward causation is logically impossible. My reason for saying that this is not possible is that the arguments which have been offered in support of the thesis that backward causation is logically impossible, and which are also compatible with a tenseless approach to time, turns out to be, at least in the first instance, arguments against causal loops, and it turns out that there is no plausible way of extending those arguments into arguments that rule out oppositely directed causal chains in worlds where causal loops are impossible.

Le Poidevin can, however, avoid this problem, since if there is a successful argument against the possibility of causal loops, that will establish that causation is asymmetric. The prospects for such an argument are, I think, not unpromising. But it would be good if Le Poidevin indicated the type of argument that he has in mind here.

7.4 The Crucial Objection

Let us turn to the second, and crucial objection to Le Poidevin's proposal. The thrust of this objection is that the assumption that causation is asymmetric does not suffice to solve the problem. In particular, it does not provide a reason for accepting (C_4), while rejecting (C_4*). To see why this is so, we need to consider my original argument, and what we'll see when we do so is that my argument is perfectly compatible with Le Poidevin's assumption that causation is asymmetric.

The second part of the argument from causation centres upon the question of how it is possible for there to be laws that satisfy postulates (C_1) through (C_4), and the thrust of the argument is that an explanation can be offered of how this is possible if and only if our world is dynamic, rather than static.

Let us begin by supposing that our world is a static world, where what facts there are does not depend upon what time it is: events are either actual or not, *simpliciter*, rather than being actual as of some times, and not as of others. Given such a world, how can one explain the existence of laws that satisfy postulates (C_1) through (C_4)? In particular, how can it be the case that the posterior probability of an effect depends upon the prior probability of the cause, in the way indicated by postulate (C_4) – namely:

$$\text{Prob}(Qx, P \Rightarrow Q \ \&\ C) = \text{Prob}(Px, C) + \text{Prob}({\sim}Px, C) \times \text{Prob}(Qx, {\sim}Px \ \&\ C)$$

whereas, as is shown by postulate (C_1)namely:

$$\text{Prob}(Px, P \Rightarrow Q \;\&\; C) = \text{Prob}(Px, C)$$

the posterior probability of a cause does not depend upon the prior probability of its effect?

If the problem were merely to explain the relationship that is expressed by postulate (C_4), it might seem that a solution was readily available. For consider any temporal slice of the world. "Prob(Px, C)" gives one the prior probability of there being an event of type P at any location x within that temporal slice. Suppose, now, that it is a causal law that events of type P causally give rise to events of type Q. It must then be the case that, whenever one has an event of type P at any location x within the one temporal slice, there must be an event of type Q at some corresponding location within a slightly later temporal slice of the world. Accordingly, given that it is a causal law that events of type P causally give rise to events of type Q, it follows that the posterior probability of an event of type Q must be at least as great as that of an event of type P. Then, by taking into account the possibility that an event of type Q may occur without being caused by an event of type P, one can derive postulate (C_4) – provided that one can assume that the posterior probability of an event of type P is equal to its prior probability.

The difficulty, however, is that there is no way of justifying the latter assumption, if the world is not dynamic. This emerges very clearly once one notices that, in a static world, a very similar, and, apparently, equally acceptable argument can be used to establish that the posterior probability of a cause depends upon the prior probability of its effect. Consider any temporal slice of the world. "Prob(Qx, C)" gives one the prior probability of there being an event of type Q at any location x within that temporal slice. Suppose, now, that it is a causal law that events of type P causally give rise to events of type Q. It must then be the case that, whenever one does *not* have an event of type Q at any location x within the one temporal slice, there cannot be an event of type P at the corresponding location within a slightly earlier temporal slice of the world. Accordingly, given that it is a causal law that events of type P causally give rise to events of type Q, it follows that the posterior probability of there not being an event of type P must be at least as great as that of there not being an event of type Q. Then, by taking into account the possibility that an event of type P may fail to occur at a certain location without there being an absence of a later event of type Q, at the appropriate location, that would be incompatible with the occurrence of an event of type P at the relevant location, and, by assuming that the posterior probability of the non-occurrence of an event of type Q is equal to the prior probability of its non-occurrence, one could show that the following analogue of postulate (C_4) must be true:

(C_4*) $\text{Prob}(\sim Px,\ P \Rightarrow Q\ \&\ C) = \text{Prob}(\sim Qx,\ C) + \text{Prob}(Qx,\ C) \times \text{Prob}(\sim Px,\ Qx\ \&\ C)$

But (C_4*) says that the posterior probability of the non-occurrence of a cause depends upon the prior probability of the non-occurrence of an effect in precisely the same way that, according to (C_4), the posterior probability of an effect depends upon the prior probability of its cause. It then follows that the posterior probability of a cause depends – albeit in a slightly different way – upon the prior probability of its effect. But such dependence is ruled out by postulate (C_1) – namely:

$$\text{Prob}(Px, P \Rightarrow Q\ \&\ C) = \text{Prob}(Px, C)$$

for the latter asserts that the relevant posterior probability of a cause must always be equal to its prior probability.

But cannot this second argument be challenged? In particular, what is the justification for the assumption that the posterior probability of the non-occurrence of an event of type Q is equal to the prior probability of its non-occurrence? The answer is: none. But equally, there is no justification – if one assumes that the world is static – for the corresponding assumption in the first argument – namely, the assumption that the posterior probability of an event of type P is equal to its prior probability.

To sum up. While there is an argument that might seem to be available, in a static world, to establish that postulate (C_4) holds, that argument cannot succeed, and this for two reasons. First, the argument involves a crucial assumption that cannot be justified in a static world. Secondly, there is a parallel argument – and one which is equally acceptable, given a static world – which can be employed to show that postulate (C_1) cannot hold. In a static world, then, it is impossible to explain how the crucial asymmetry expressed by postulates (C_1) and (C_4) can obtain" (*Time, Tense, and Causation*, 107-110).

The fundamental objection to Le Poidevin's proposal will now be clear. The assumption that causation is asymmetric means that it cannot be true, for example, that both $P \Rightarrow Q$ and $Q \Rightarrow P$ are causal laws. The above argument, however, does not assume that this is the case. It involves, on the contrary, only the one causal law – $P \Rightarrow Q$, and the problem arises because that one causal law entails the following two nomological statements:

(1) It is nomologically necessary (x)(Px ¢ Qx);
(2) It is nomologically necessary (x)(~Qx ¢ ~Px).

The problem then is that while statement (1) provides the basis for an argument that justifies postulate (C_4), statement (2) provides an equally good basis for an argument that justifies postulate (C_4*).

7.5 The Popper Case: How Does One Explain the Asymmetry?

Finally, let me conclude by describing a consideration that makes vivid, I think, the appeal of the line of argument that I have offered in support of a dynamic world. It involves the situations to which Popper appealed in offering an account of the direction of time that was based upon transmission of order outward from a central cause – the sort of situation that exists when a stone is thrown into a pond, giving rise to outgoing concentric waves, or that exists when a light is turned on, resulting in outgoing spherical wave fronts of light. Popper's point was that such processes occur very frequently indeed, whereas the flipped over processes – where, for example, light waves converge at the same time to a central source, where they are all absorbed – occur never or virtually never. Thus Popper proposed that one could identify the direction of time with the direction of such processes.

Suppose, now, for simplicity, that our world is a Newtonian world, and compare the following two situations:

Situation 1 A stone is moving towards a pond that is calm, and without any waves;
Situation 2 A stone is moving away from a pond that is calm, and without any waves.

Situation 2 differs from situation 1 only in that the direction of the motion of the stone is reversed. Consequently, the prior probability of the two situations should not differ. Why then is it that situation 1 occurs frequently, whereas situation 2 occurs virtually never? If postulates (C_1) through (C_4) are true, an answer is available. For when one applies the Newtonian laws of motion to situations 1 and 2, one can see that the cause of situation 2 would be an extremely complex event, involving incoming concentric waves on the pond, whose prior probability is incredibly low, whereas this is not so in the case of situation 1. But, then, given postulates (C_1) through (C_4), it turns out that the posterior probability of situation 2 must also be extremely low, whereas that of situation 1 is not.

This argument presupposes, however, that one cannot use a "backtracking" argument, involving (C_4*), in order to argue that, given the prior probability of situation 2 is not especially low, the posterior probability of the earlier, very complex event involving incoming concentric waves on the pond, must also not be especially low. But this "backtracking" argument will be just as reasonable as the "forwardtracking" argument if the tenseless view of time is correct. It is only on a dynamic view of time – and, specifically, only on a dynamic view according to which the past is real, while the future is not – that one can justify *both* using the forwardtracking argument, while rejecting the backtracking argument, and thus explain why it is the case, in our world, that situation 1 is immensely common, whereas situation 2 occurs never, or virtually never.

SUMMING UP

A full defense of the Argument from Causation would involve arguing for the plausibility of the realist account of causation upon which the argument rests. I have not attempted to do that in the present paper. But I have argued elsewhere both that there are decisive objections to reductionist approaches to causation, and that the realist account that I have offered can be supported by a variety of positive arguments.

What I have attempted to do in the present paper, is, first, to show that the Argument from Causation escapes Le Poidevin's objections, and, secondly, to describe one of the important advantages that result from combining an account of causation that rests upon postulates (C_1) through (C_4) with a dynamic view of the nature of time according to which the past and the present are real, but the future is not.

24
REPLY TO SMITH AND TOOLEY

Robin Le Poidevin
University of Leeds

I am very grateful to Quentin Smith and Michael Tooley for their very thoughtful comments on my paper. They raise many issues, and I shall only attempt to address some of them here, and augment my original argument at one or two points. First, it may be helpful if I clarify the strategy of my paper.

The relation of temporal precedence is under attack from two completely opposite directions. If the tensed theory of time is true, then more fundamental than facts about precedence are tensed facts: facts about the pastness, presentness or futurity of events. Precedence would then supervene on tensed facts in the way that physicalists hold mental facts to supervene on physical ones. Precedence would then be a merely secondary quality in one sense of the term. On the other hand, if the tenseless theory is true, then there is an argument according to which the asymmetry of causation is not intrinsic to the causal relation, but rather a feature of our perspective on the world (just as, according to the tenseless theorist, tense is a feature of our perspective on the world).[1] If that argument is sound, then we have a further argument to the effect that temporal precedence is similarly a feature of our perspective on the world. The argument, in a nutshell, goes like this: either temporal precedence depends on causal precedence or it is independent of it. If the relation is one of dependence, then any proof that causal asymmetry is purely perspectival will generate the conclusion that temporal asymmetry is similarly perspectival. If temporal precedence is independent, then it would provide a way of distinguishing cause from effect, and so undermine the argument for causal perspectivalism. Causal perspectivalism, then, goes hand in hand with perspectivalism about precedence, which then would be merely a secondary quality in the Lockean sense. Assuming we allow time to be real, then some version of either the tensed or tenseless theory will be true, so there is reason to suppose that precedence is, after all, a secondary quality (in some sense).

It is this conclusion I wanted to resist. I could see no plausible way of saving precedence from relegation to secondary status if the tensed theory were true, so my attention was focused on the tenseless theory. As a tenseless theorist myself, I find causal perspectivalism particularly unwelcome,

L. N. Oaklander (ed.), The Importance of Time, 285–291.

as it undermines the tenseless theorist's best explanation of the perception of time. The argument I considered at the beginning of the paper was an attempt to make this absolutely explicit. The crucial premise of that argument was

(3) The objectivity of causal asymmetry entails that the future is open

and it is this premise I shall be mainly concerned with in this reply. It is not the most transparent of premises. In particular, it needs to be made clear just what sense the future is, or could be, open. I identified two interpretations:

(i) *token* indeterminacy
(ii) *type* indeterminacy

Since Tooley does not regard his own account of the unreality of the future as explicable in either of the above ways, we should acknowledge a third:

(iii) non-actuality as of earlier times

To add a further level of complexity: "earlier times" can be characterized either in tensed terms or (as in Tooley's case) tenselessly.

Now let us take a look at the entailment in (3) in the light of these different interpretations of the open future. If objective causal asymmetry does indeed entail an open future, what explains this logical relation? There are two possibilities:

(a) The analysis of the causal relation requires the openness of the future.
(b) Some characteristic feature of causation (which yet is not part of the analysis) can only be explained by the openness of the future.

Let us take the first of these. Bearing in mind the different interpretations of the open future, what (a) implies is that causation is defined in terms of either (i) making token determinate; or (ii) making type determinate; or (iii) making actual. If we opt for (i) then we are contemplating a picture according to which a cause determines, not only that an effect of such-and-such a type will occur, but *that very token* of that type will occur. I find this notion unintelligible, but both Smith and Tooley disagree and provide putative examples of such token determinacy. Smith's example is of a universe consisting entirely of three electrons, which exist throughout the history of the universe. Given determinism, the state of every token individual is fixed for every time. However, although I do not disagree with Smith's description of the case, I don't think it will do the job, as the token states of this universe – that is, the obtaining of token relations between the electrons – will not be determined, and it is these states that are the effects in question. So the future, conceived of as a series of states, will not be *made* token determinate by the past (even it is as a matter of fact token determinate). Similar remarks apply to Tooley's example of a single electron. It may well be that conservation laws entail the token identity of an electron existing at some future time,[2] but neither they nor anything else will determine the token state of that electron.

So what if we take causation as defined in terms of making type determinate? This, I suggested, was incompatible with determinism, for determinism implies that all states of the universe are type determinate for all time, and not just when their immediate causes occur. That in itself is an unwelcome consequence. But perhaps we can push the problem further. If causes make type determinate their effects, does this not imply that causation is, after all, deterministic? If so, then the analysis of causation as making type determinate faces paradox: it implies that the universe is, and is not, deterministic. We could try to avoid paradox here by characterizing determinism on the one hand as fundamentally a *modal* notion (thus the state of the universe at *t* plus the laws of nature logically necessitate a certain series of states at other times), and characterizing the relation of making type determinate on the other as *non*-modal (so that it is purely contingent that this cause gives rise to that effect), but the worry still remains that if the type determinacy of the effect is to be attributed to the cause, then it becomes determinate (though not yet existent) when the cause occurs. But then if its determinacy does indeed precede its existence, then it does so at a time when its remote causes come into existence. In other words, there is never a time when the effect is not type determinate, so defining causation as making type determinate undermines this way of charactering the open future.

Finally, what of causation as making actual? This is not Tooley's conception of causation, but it does borrow his notion of being actual-as-of-a-time. If that notion turns out to be incoherent, then of course we cannot base an analysis of causation upon it. So what I propose to do now, at some length, is to examine the coherence of Tooley's conception of the non-actuality of the future.

In my view, if it is a genuinely metaphysical and not just epistemological matter, the putative openness of the future, whatever interpretation you give it, entails the existence of tensed facts. This is one point on which Tooley and I disagree. For Tooley, the openness, or unreality, of the future can be characterized entirely in terms of tenseless facts: in terms of what is actual at a (tenseless) time. Now whatever actuality at a time is, it must be more than existence at a time. Tensed and tenseless theorists alike agree that states of affairs exist at some times and not others. The issue is what the status of states of affairs is at times when they do not exist. Now, for Tooley, the status of a state after it has ceased to exist is very different from its status before it begins to exist. It remains part of what is actual even when it has ceased to exist, but is not yet actual before it begins to exist. But this difference in status is not supposed by Tooley to be absolute, but rather relative to a time. That is, a state can be actual as of t_1 but not actual as of t_0, where t_0 and t_1 are to be characterized in purely tenseless terms. Now one way of characterizing actuality at a time is to say that if a state is actual at a time, then there are truths about that state at that time (even if the state *exists* at an earlier time). Conversely, not being actual at a time means that there are no truths about that state at that time. It is true at t_2, for example, that the state at t_1 included a noisy demonstration outside the White House. It was not true at t_0, however, that the state of affairs at t_1 included that noisy demonstration, as the state is not actual at t_0. But is it coherent to say that that are truths about a state after it exists, but no truths about it before it exists? Tooley thinks so, and offers the following in defence:

> But given that *s* is an instantaneous state of affairs that exists at time t_1, the only thing that can make it true at time t_1 that *s* exists is something – namely *s* – that exists at time t_1. This, however, will not be part of what is actual as of the earlier time t_0, and so what is actual as of time t_0 cannot contain any truthmaker for the proposition that it is true at time t_1 that *s* exists. (p. 272)

Now I take it that there is no assumption here that the truthmaker for any proposition concerning what is true at a time t must itself *exist* at t. Tooley would, I am sure, agree that a state *s* existing at t_0 can provide the truthmaker for the proposition that at a later time t_1 it is true that s exists at t_0. What state of affairs exists at t_1 is irrelevant. (Only the traditional tensed theorists would deny that.) But if truthmakers can exist at times other than the time at which truths about them are framed, then why cannot some propositions have *later* truthmakers? Tooley would say that later truthmakers are not actual, but this looks suspiciously like stipulation.

Let me elaborate. Consider a very simple universe: one consisting of three times, and three states obtaining at those times. Now consider the description of that universe from the standpoint of three theories of time: the traditional tensed theory, which explains the unreality of the future in terms of tensed facts; the traditional tenseless theory, and Tooley's own account:

(D1) *Description of reality according to the traditional tensed picture*

Then: s_0 exists now.
Now: s_0 existed earlier. s_1 exists now.
Later: s_0 existed earlier. s_1 existed earlier. s_2 exists now.

Tensed facts change over time. The past is real, the future unreal. No truth *simpliciter*, just truth-at-a-time.

(D2) *Description of reality according to the traditional tenseless picture*:

t_0: s_0 exists at t_0. s_1 exists at t_1. s_2 exists at t_2.
t_1: s_0 exists at t_0. s_1 exists at t_1. s_2 exists at t_2.
t_2: s_0 exists at t_0. s_1 exists at t_1. s_2 exists at t_2.

Tenseless facts do not change over time. All times equally real. No truth-at-at-time, just truth *simpliciter*.

(D3) *Description of reality according to Tooley's picture*:

t_0: s_0 exists at t_0.
t_1: s_0 exists at t_0. s_1 exists at t_1.
t_2: s_0 exists at t_0. s_1 exists at t_1. s_2 exists at t_2.

Tenseless facts change over time. Earlier times are real, later times unreal. Both truth-at-a-time *and* truth *simpliciter*.

D1 is a description of reality from a particular temporal perspective. It is not possible, on the traditional tensed account, to give a description that is not from a temporal perspective. Both truths and their truthmakers change over time, so it is clear why the truthmakers of a statement made *now* must itself exist *now*. D2 is a description of reality from no particular perspective. Neither truths nor truthmakers change over time, so there is no reason to stipulate that truthmakers and the items they make true should stand in any particular temporal relation to each other. Now is there any room in logical space for D3? Let us ask first whether it is a description of the world from a particular perspective, or from no particular temporal perspective. Tooley himself insists that a satisfactory account of time requires both the notion of actuality-at-a-time and of the totality of existence: actuality at some time or other.[3] D3, then, we may imagine to be a description of the totality of existence, and so true of the world *simpliciter*. What, then, are its truthmakers? Well, presumably, s_0, s_1 and s_2. But given the timeless reality of these truthmakers, why are they debarred from making true any proposition about what facts there are at earlier times? In other words, there is at least a tension between allowing truth *simpliciter* and insisting that facts at one time are not actual at earlier times. So what if we resolve the tension by simply jettisoning the notion of truth *simpliciter*? Then of course D3 becomes a description of the world from a temporal perspective, and is only available to us as a description of the world if we are at t_3. So what is true is what is *now* true (from our current perspective), and to reflect the ontological significance of our current perspective we have to introduce tensed facts – the fact for instance that D3 *is now* a correct description of reality. D3, then, is unstable, and collapses either into D1 or D2.[4]

Of course, this does not entirely dispose of the notion of the non-actuality of the future, but it does require us to view it in tensed terms. The future is not merely non-actual relative to a tenseless time, but is non-actual, if at all, in an absolute sense. So where does this leave us? We began by looking at the premise

(3) The objectivity of causal asymmetry entails that the future is open

and asked what explained the relation of entailment. One answer was that the analysis of the causal relation requires the future to be open. Plugging in the various conception of the open future, we have the following possible analyses of causation: as (i) making token determinate; (ii) making type determinate; (iii) making actual in either a tensed or tenseless sense. (i) we ruled out as unintelligible. (ii) we ruled out on the grounds that it committed one to indeterminism, and possibly resulted in paradox. (iii) we argued could not be coherently characterized in a purely tenseless way. So the only conception of causation we are left with is that of making actual, where being actual is defined neither in terms of token-determinacy, nor of type determinacy, nor in purely tenseless terms. Now if being actual is to be explicated neither in terms of being existent, nor in terms of being type or token determinate, it seems we must regard it as a primitive notion. If so, then I have no argument against the suggestion that causation is to be analysed in terms of the primitive notion of making actual, but it does not strike me as a particularly enlightening or compelling suggestion to make, and so rather inadequate grounds for establishing the truth of (3).

Even if the openness of the future is not required by the analysis of causation, however, it may be required to explain some characteristic feature of causation. This is Tooley's position, and he argues for it terms of certain probabilistic postulates. The argument is well summarized in Tooley's response to my paper, so I will not repeat it here. I will, however, try to clarify my response in the light of a remark of Smith's. Smith is unsure of my grounds for the assertion that the asymmetry of causation itself is the basis for the fact that the probability of the effect depends on that of the cause in a way in which the probability of the cause is not dependent on the probability of the effect. It is true that I was not at all explicit about these grounds in my paper, although Smith does provide a good partial reconstruction of them. The reasoning was this: given the failure of causal perspectivalism, we are justified in our assumption that causation is objectively asymmetric. Further, there are no compelling reasons to build the unreality of the future into the analysis of causal asymmetry. It is reasonable, then, to try to account for any asymmetry between cause and effect, such as the asymmetry in their probabilistic relations, in terms of the logical asymmetry of causation itself, rather than in terms of the supposed ontological asymmetry between past and future. Indeed, one would expect any asymmetry between cause and effect to arise from the most fundamental asymmetry between them, rather than from some quite separate feature of the world.

Tooley however objects that the asymmetry of causation does not explain the probabilistic asymmetry he has identified. Let me make two points in reply. First, suppose that he is right. Given what we have just said about asymmetries of cause and effect flowing from some fundamental asymmetry of the causal relation, we might then be suspicious of the suggestion that there is, after all, a probabilistic asymmetry between cause and effect, and therefore of the suggestion that there is something to be explained. Second, we need to be clear exactly what kind of probabilistic relation we have in mind. If it is merely probabilistic *covariance*, then we would expect the probability of the cause (or non-occurrence of the cause) to vary with the probability of the effect (or non-occurrence of the effect), even given the asymmetry of the causal relation. But if we are considering probabilistic *dependence*, then this looks much more like a genuinely causal relation, or one that would arise from the causal relation.

NOTES

[1] I explored part of the motivation for this position in the original paper.
[2] Even this could be questioned, however, if we regard the electron through time, not as a continuant, but as a mereological sum of temporal parts. The token identity of one temporal part would not determine the identity of a later temporal part.
[3] See Tooley (1997, p. 41).
[4] For a related criticism of Tooley's account see Oaklander (1999), and Essay 1 of this volume.

REFERENCES

Oaklander, L. N. (1999), "Review of *Time, Tense, and Causation by Michael Tooley*," in *Mind* 108: 407-413.
Tooley, M. (1997), *Time, Tense, and Causation* (Oxford: Clarendon Press).

NAME INDEX

MIX
Papier aus verantwortungsvollen Quellen
Paper from responsible sources
FSC® C105338

If you have any concerns about our products,
you can contact us on
ProductSafety@springernature.com

In case Publisher is established outside the EU,
the EU authorized representative is:
Springer Nature Customer Service Center GmbH
Europaplatz 3, 69115 Heidelberg, Germany

Printed by Libri Plureos GmbH
in Hamburg, Germany